HISTORY OF THE CLIMATE CHANGE ON THE COROMANDEL COAST NINTH-NINETEENTH CENTURIES

History of the Climate Change on the Coromandel Coast

Ninth–Nineteenth Centuries

S. JEYASEELA STEPHEN

MANOHAR
2023

First published 2023

This edition is for sale in India, Sri Lanka, Nepal, Bangladesh, Afghanistan, Pakistan and Bhutan.

ISBN 978-93-94262-55-3

Published by
Ajay Kumar Jain *for*
Manohar Publishers & Distributors
4753/23 Ansari Road, Daryaganj
New Delhi 110 002

Typeset by
Kohli Print
Delhi 110 051

Printed at
Replika Press Pvt. Ltd.

Contents

List of Tables 7

Acknowledgements 9

Abbreviations 11

1. The Historical Setting 13

2. Waterscapes: The Rainfall in Tamil Country, Ninth-Nineteenth Centuries 22

3. The Famine and Drought in Tamil Society, Ninth-Nineteenth Centuries 67

4. The Storms and Cyclones of Tamil Littoral and the Europeans, Seventeenth-Nineteenth Centuries 98

5. Hazards of Sea, Land and Water: Floods, Tsunamis and Earthquakes, Ninth-Nineteenth Centuries 127

6. The Study of Temperature and Atmospheric Pressure: Technology Transfer from Europe to Tamil Coast, Eighteenth-Nineteenth Centuries 145

7. Concluding Remarks 174

Appendices

Appendix I 185

Appendix II 186

Appendix III 188

Glossary 191

Bibliography 197

Index 211

List of Tables

2.1. Rainfall Estimates in Madras in Millimetre for 1733-6 39
2.2 Rainfall in Pondicherry, 1761-2 41
2.3. Rainfall in Pondicherry, 1794 41
2.4. Rainfall in Pondicherry, 1795 42
2.5. The Mean Fall of Rain in Inches at Madras for the years between 1803 and 1819 43
2.6. The Mean Monthly Fall of Rain at Madras Including the Fall during the Storms in Eighteen Years, 1803-25 44
2.7. Rainfall in Madras in Days and Inches, 1827-76 46
2.8. Daily Rainfall of Given Amount in Inches in Each Month of the Year during 1875-87 in Madras 47
2.9. The Mean Rainfall on Drainage Basins of Rivers, 1817-76 50
2.10. Rainfall in Tiruchirapalli in Days and in Inches, 1827-76 50
2.11. Rainfall in Madurai in Days and in Inches, 1827-76 51
2.12. Rainfall in Coimbatore in Days and in Inches, 1827-76 51
2.13. The Fall of Rains in Inches at Palayamkottai and Shenkottai Reported by General Cullen, 1844-8 52
2.14. Average Monthly Rainfall in Inches at Madras, Cuddalore, Tiruchirapalli, and Madurai for 1875-85 54
3.1. The Periodicity and Frequency of Famines in the Localities of Medieval Tamil Country, AD 1019-1396 69
3.2. Famine Years in Coromandel, AD 1520-40 74

3.3. Famine Years in Madras, AD 1646-1779 77
3.4. Famine Years in Thanjavur, AD 1743-1830 81
3.5. Famine Years in Tiruchirapalli, AD 1805-65 82
3.6. Famine Years in Madurai Region, AD 1804-99 82
3.7. Famine Years in Madras, AD 1812-97 83
4.1. The List of Storms Originated from the Various Ports on the Coromandel Coast, 1842-87 112
4.2. Register of the Sympiesometer, Barometer, and Force of Wind at Madras, 25 November 1846-26 November 1846 taken by J.J. Franklin 121
4.3. Daily Movement of the Wind in Madras and Nagapattinam in 1885 121
5.1. The Periodicity and Frequency of Floods in the Localities of Medieval Tamil Country, AD 937-1402 128
5.2. Earthquakes in Tamil Country, 1807-1900 140
6.1. The Mean Annual Heights of the Barometer and Thermometer (with the General Mean of Each) Shown at Madras, AD 1796-1821 153
6.2. The Mean Monthly Heights of the Barometer and Thermometer at Madras, AD 1796-1821 154
6.3. The Mean Monthly Dry of the Hygrometer at Madras, AD 1819-23 155
6.4. Mean Height of the Thermometer and Barometer during the North-East Monsoons at Madras between AD 1796 and 1821 156
6.5. Mean Height of the Thermometer and Barometer during the South-West Monsoons between AD 1796 and 1821 156
6.6. The Highest Temperature Registered with Thermometer for Each Month at Madras in 1885 and the Highest Temperature in the Day 164
6.7. Mean Daily Range of the Thermometer in Ootacamund, Wellington and Coimbatore in 1885 167
7.1. Famine Years with Less Annual Rainfall (in 10th of mm) between 1871 and 1900 176
7.2. Flood Years with Heavy Annual Rainfall (in 10th of mm) between 1872 and 1898 178

Acknowledgements

The book is part of the ongoing personal research on the scientific history of the Tamil coast. It is a continuation of research and analysis of the histories that were constrained by the Tamil coast and hinterland, and the historical impact of forces that were operating at the regional, continental or global scale by investigating into themes, trends and relationships. The study of botany, chemistry, medicine, earth science, space science and animal science has been published in two monographs. This is the third book in this series and the theme of history of climate was chosen and 'water', a topic of great importance began to figure, particularly as found in the Tamil maritime history, with the sea coast and its long shorelines in relation to its landmass, and with the enormous expanse of the hinterland. Water, in several respects a necessity of life, is at the same time a very dangerous element owing to floods and, in the absence of rain, the trouble it causes to people in society, thus constituting a kaleidoscopic image of climate history. It is hoped that the advantages of the novel approach chosen here would be welcomed by scholars of environmental history also by making the longue durée (long time) approach in history from the ninth to nineteenth centuries. The changes and continuities in the pre-modern and modern age have been aimed to be investigated in this volume.

I would like convey my sincere thanks to those institutions that have funded my earlier projects that helped in making this book possible. Thanks are due to the many archives, libraries and repositories that facilitated in collecting data used in this book, particularly Archiv der Franckesche Stiftungen, Halle; Nationaal Archief, Den Haag; British Library, London; Archives Nationales, Paris; Archives de la Societe des Missions Etrangeres de Paris, Paris; The Asiatic Society, Kolkata; Tamil Nadu State Archives, Chennai; National Archives of India, Regional Record Centre, Puducherry; Regional Meteorological Centre, Chennai; Indian Institute of

Tropical Meteorology, Pune; Meteorological Department, Chennai; and the library of the French Institute, Puducherry.

Many scholars have given constant encouragement and extensive suggestions that proved valuable in the course of writing of this book. Sincere thanks to Prof. Salvatore Ciriacono at the Universita degli Studi di Padova, and Prof. Martin Krieger at the Christian-Albrechts-University of Kiel. I extend my gratitude to Shri Ramesh Jain and Shri Ajay Jain of Manohar Publishers for undertaking the publication of this book. I hope this book on the history of weather and climate helps to advance considerably our knowledge.

Puducherry S. JEYASEELA STEPHEN
12 June 2022

Abbreviations

AFSt	Archiv der Franckesche Stiftungen, Halle
AMEP	Archives de la Societe des Missions Etrangeres de Paris, Paris
AN	Archives Nationales, Paris
ARE	Annual Report on Epigraphy
BAA	Bulletin Actes Administratif
BL	British Library, London
EFI	English Factories in India
EI	Epigraphia Indica
FRS	Fellow of the Royal Society
HB	Hallesche Berichte
IPS	Inscriptions of Pudukottai State
LEC	Lettres Edifiantes et Curieuses
MS Eur	European Manuscript
NA	Nationaal Archief, Den Haag
NAIP	National Archives of India, Regional Record Centre, Puducherry
NHB	Neue Hallesche Berichte
OIOC	Oriental and India Office Collection
PWD	Public Works Department
SII	South Indian Inscriptions
SITI	South Indian Temple Inscriptions
SPCK	Society for the Promotion of Christian Knowledge
TAS	Travancore Archaeological Series
TNSA	Tamil Nadu State Archives, Chennai
TTDES	Tirumala-Tirupati Devasthanam Epigraphical Series
VOC	Verenigde Oost–Indische Compagnie

CHAPTER 1

The Historical Setting

In the course of writing the history of south India, scholars in the beginning, traced its historical geography. They limited their scrutiny to the location of places, boundaries of empires and routes of the marching armies. K.A. Nilakanta Sastri (1892-1975) made an exceptional contribution, formally expounding the influence of geographical factors on historical development.[1] Burton Stein (1926-96) recognized the influence of ecological factors on the agrarian and human aspects in the Tamil region.[2] In the study of historical geography based on the Chola inscriptional records, two methods were employed for handling heterogeneous spatial data in a chronological context. The first method focused attention on specific elements of change, and clear selections were made around which to organize the material. The second method was that the past geography at particular points in time had been reconstructed with emphasis on the spatial relationship with less attention to the processes of change.[3] Thus historical geographers have contributed significantly to the development of environmental history.

TAMIL LANDSCAPE

According to Sangam literary sources, there were five types of *thinais* (landscapes) in ancient Tamil country. The *kanavar, kunavar* and *vedar* were the people of the *kurinji* (hilly region); the *aayar* and *idaiyar* belonged to the *mullai* (forest region); the *uzhavar* and *thozhuvar* lived in *marutham* (wetlands); the *parathavar, valayar* and *meenavar* were the people of *neithal* (coastal region); and the *kalavar, eyinar* and *maravar* were of *paalai* (desert area).[4] Some writers have opined that these physiographical divisions cannot be treated as geographical because they merged with one another and their exact demarcation is impossible.[5] However, it is important to note that there cannot be watertight compartments of landscapes and people lived in various zones in the beginnings of human civilization.

Many rivers flow in the Tamil region and the Kaviri is an important river. In Tamil, the term 'Kaviri' means 'the river which enriched the gardens'.[6] Also called Kaveri, this river has been considered as very holy, and bathing in it or performing penance on its banks was believed to be efficacious. The Kaveri was said to be the lifeline of the Chola kingdom and it has been extolled in the epic Ramayana.[7]

Tamil texts mention rivers in a poetic vein. The Tiruchirapalli rock inscription of Mahendravarman (600-30) says that the Kaveri was very pleasing to the eye.[8] The Kanyakumari inscription states that River Kaveri carried on earth, in the form of water, the very nectar obtained by the Gods from the ocean of milk.[9] Udaya Chandran, a vassal of Nandivarman II Pallavamalla (731-96) is described as the Lord of River Vegavati.[10] Kopperunjika I, the Kadva ruler (1231-63), bore the title 'Kaveri Kamukha' (the lover of Kaveri).[11] River Vaigai, another important river, has been prefaced with the description of the flooded Vaigai with the scene of downpour of rain in the Saiyam hills.[12] Thus, rulers had realized the importance of the rivers and it is interesting to note that water levers which helped to transfer the force of water to the load providing mechanical advantage had been assigned while making land grants by the rulers.[13]

The economic prosperity of the Tamil kingdoms depended to a great extent on water and rains. Agriculture was chiefly dependent on a sound irrigation system. Thiruvalluvar in his *Thirukkural* (also called *Kural*) in the ancient period wrote 'if the heavens dry up, the very gods will lack festival and worship'.[14] He also said 'should rain fail, hunger will rack the wide earth sea-grit'.[15]

EVENTS OF FLOODS IN ANCIENT SEAS

The subcontinental landmass known as 'Terra Tamilica' in ancient times was located near the large island then called Naavalam or

Jambhu Thivu as mentioned in the Tamil epic poem *Manimekalai.* According to this early Tamil literary text, part of the port of Kaviripoompattinam was swallowed by the angry sea.[16] Terra Tamilica was one of the earliest known land formations to take shape on either side of the equator. *Silapathikaram* and *Pathittru Paththu* of the same period cite that the prosperity of the Tamil land owed its advantage chiefly to its geographical location, surrounded by the sea and the ocean.[17] The submerging of Kaviripoompattinam in the ocean is an instance of the devouring of land by sea in the early Christian era. The inundations that took place later cut off Sri Lanka from the Tamil land. The Pandya capital of Kavatapuram or Muthur, the port noted for its pearl fishery, met with the same fate.[18] Thus, an ancient Tamil continent was lost to the sea in a series of great deluges.

These floods mentioned in the Sangam literature had been called myths by some scholars. They relate the events mentioned above, thus, once the sea rose against the ancient city of Madurai; the Gods were alarmed and, seeing this, Siva appeared to Ukkira Pandyan in a dream and told him to throw his lance which he gave. The Pandyan awoke and, after being urged by the God, threw his spear at the sea, which became calm and lapped at his feet.[19] The event narrated in the Tamil Sangam literature thus ascribed flood to the divine cause.

MEDIEVAL COASTAL INUNDATIONS

References to coastal inundations had continued to be mentioned in medieval inscriptions. A series of disasters have also been recorded. A severe sea inundation occurred in AD 1124 at Tiruvaththur.[20] Sources indicate that due to such inundations from the sea, land was submerged and silted up in Sirkazhi area. Such lands were reclaimed by paying 500 *kasu* per *maa* of land.[21] History of the Tamils thus involved humans interacting with the environment.

Tamil historians have not focused adequately on the interactions between humans and nature. Many writings have tended to do a variety of traditional political history. The knowledge of past environments had not adequately grown by studying nature through history and by studying history through nature. Historians of science and historians of ideas in the modern age have written mainly about the development of ideas of nature in science and also produced works of environmental history. Philosophers have often displayed a conscious tendency to separate human history, as R.G. Collingwood (1889-1943) pointed out—from natural history, sometimes proceeding even to deny that nature could ever have history quite in the same way humans have it. Hence, the need of the hour is researching on climate history, which is the study of interactions between both extreme weather events and normal as well as climatic shifts and their impacts on Tamil society. A great strength of this book is the collaboration between history and science.

SPATIAL AND TEMPORAL ASPECTS OF THE STUDY

The study area chosen for this book is the Coromandel Coast, the Tamil littoral on the east coast of the Indian peninsular. The Dutch and the English extended the region in the age of expanding their commerce at the close of the seventeenth century up to the south of the Godavari river in the Telugu-speaking region. One must be careful in using the term 'Presidency of Madras', which was newly coined and came to be widely used in English Company records to suit the purpose of the administration. The Madras Presidency's extent covered the whole of Tamil region besides the northern part of modern Kerala (the Malabar), coastal Andhra Pradesh (Rayalseema), Berhampur and Ganjam districts of modern Odisha (earlier known as Orissa) and Bellary district of modern Karnataka. Climatic variance was totally different in all these zones. Therefore, this present study area of Tamil country is limited only to the Tamil-speaking region.

SOURCES AND METHODS

The sources used are medieval Tamil inscriptions, literary sources and the accounts of European travellers, missionaries, and the servants and officials of the East India Company. Portuguese, Dutch, Dane, German, French and English traders and missionaries who arrived on the Tamil coast and saw the weather and climate totally different from their countries in Europe. Hence, they began to study the climate on the Tamil coast. They reported the social and environmental impacts of natural hazards on the Tamil coast, especially the seas, storms, floods, tsunamis, earthquakes and droughts that destroyed substantial human and economic resources, and stood as formidable barriers against regional development at regular intervals. By examining these records on the weather and climate, it is reasonable to suppose that some useful lessons might be learnt from past events especially from the study of the impact of climatic variation and conditions which have also been experienced in recent times. Therefore, the book adopts the methods of statistics, analysis and comparison.

STRUCTURE OF THE STUDY

Chapter 1 introduces the subject of climate history of the Tamil landscape and explains how environmental historical studies done in the past are different from the study of climate history. It outlines the significance of multidisciplinary field of research, linking historic climatic studies by demonstrating how hazards can be considered both as events and in the case of frequent disasters as processes.

Chapter 2 presents an account of rainfall of 1,100 years and looks into how changes in physical climates have influenced Tamil society. The study is concerned not only to show the changeability of physical climate and the adaptiveness of society to such change, but also the way people in the medieval and early modern age thought about and made sense of rainfall and its variations—the 'behaviour' of climate. The chapter offers to explain the dams built across rivers; excavation of various lakes; setting up of committees for maintenance; the role of the village headmen and the local officers in irrigation; building several canals, water channels, water tanks and sluices, digging wells; and the system of water management under the rulers. With the advent of the Europeans, the science of reading the rain had particularly developed since

AD 1732 and by measuring the rainfall through rain gauge, the Europeans collected statistical data of rainfall at various places in Tamil country and showed the gradual change, its causes and its consequences.

Chapter 3 explores how famine and drought broke out with the failure of monsoon and in the absence of proper and adequate rainfall, how food-grain collection and storage were undertaken besides the voluntary and involuntary migrations that happened to cope with the variations of shifting and anomalous patterns of drought, and crop failures. The chapter highlights how prices escalated, people entered into slavery, raised cash loans from the temple treasuries, and how the famines that came now and then disturbed the peaceful life of the people. The European traders in this period carried on slave trade. The European missionaries did religious conversions. The study traces the efforts of the colonial states, i.e. the British in Madras (now known as Chennai) and the French in Pondicherry (now known as Puducherry) in solving food-grain scarcity and rendering relief to the people and appointed the famine commission to know the facts. An analysis of climate and society interaction has been undertaken in this chapter and the study probes the famine policy of the government and the response of the population during the Great Famine of AD 1876-7.

Chapter 4 deals with the storms, hurricanes and cyclones of the Coromandel Coast and the winds generally accompanied by variable intervals of calms and furious storms since AD 1640 in Pulicat, Madras, Mylapore, Sadurangapattinam, Pondicherry, Cuddalore, Porto Novo, Karaikal, Tranquebar and Nagapattinam, in addition to the damages caused and the sufferings of the people. Lutheran mis-sionaries like Johann Ernest Geister in Madras used scientific instruments. Europeans like John Goldingham made a scientific study about the moon's impact on gales and storms from AD 1797-1820. J.J. Franklin studied the storms in 1846. Thus, this chapter uses all these data to know past climate.

Chapter 5 dwells on the floods of the Coromandel Coast between AD 937 and 1898 that occurred during the heavy fall in the rainy season during the months of October and November. These unprecedented floods affected the lives of the common people and so the role of the medieval state has been studied. It also traces floods since AD 1681 at Nagapattinam, Madras, Tranquebar, Pondicherry, Cuddalore, Thanjavur and the role of the British and the French in coming to the aid of the people through flood relief. The chapter also traces the tsunamis that occurred since AD 1788 and earthquakes that occurred since AD 1725.

Chapter 6 describes how the Europeans who came to the Tamil coast tried to make sense of the natural world. William Roxburgh, Colonel James Capper and John Goldingham developed interest in using scientific instruments like the Nairne and Blunt thermometer and the Jesse Ramsden barometer to take measurements of the changes in both temperature and sea level. William Gilchrist in Madras invented the Self-registering Barometer as well as the Metallic Tube Barometer in AD 1837. Climate change detection was made possible at the Madras Observatory through the technological enhancement and use of enhanced tools brought from Europe. Thus, this chapter examines after the advent of the Europeans the interest in climate history and the active efforts of various individuals in deciphering the Tamil region's climate and the growing interest in documenting the history of the Tamil landscape. The reports of climate at Madras, Tiruchirapalli, Madurai, Coimbatore and Nilgiris prepared by the Europeans have been analysed.

The concluding remarks dwell on comparisons of the weather and climate before and after the advent of the Europeans on the Coromandel Coast. A multilingual and global approach has been undertaken to recover the long-term records that shed light on the uniqueness of the rapid changes we are seeing today. The term 'global warming' that refers to the long-term trend in rising temperatures has been taken into consideration here. Again, the term 'climate change' that describes, more broadly, the changing precipitation and increasing the likelihood of more severe storms and hurricanes and flooding in some areas and more extreme drought in other areas in Tamil history have been undertaken. Thus, the most destructive natural hazards have been viewed in terms of their occurrence, intensity and the hinterland area affected. The impact of climatic fluctuation and change on the Tamil coast has been thus recognized.

NOTES

1. K. A. Nilakanta Sastri, *A History of South India from pre-Historic times to the Fall of Vijayanagar*, Delhi: Oxford University Press, 1976, pp. 34-6.
2. Burton Stein, 'Circulation and Historical Geography of Tamil Country', in *Journal of Asian Studies*, vol. XXXVII, no. 1, November 1977, pp. 7-26.
3. B. Suresh, *Historical and Cultural Geography and Ethnography of South India from the Chola Records*, PhD Dissertation, Poona: Deccan College, 1965.
4. K. Siva Thamby, 'Early South Indian Society and Economy: The Tinai Concept', in *Social Scientist*, vol. 29, 1974, pp. 20-37.
5. Rajan Gurukkal, 'Forms and Production of Forces of Change in Ancient Tamil Society', in *Studies in History*, vol. V, no. 2, 1989, pp. 164-5.
6. N. Subramanian, *Pre-Pallavan Tamil Index*, Madras: Ennes Publications, 1990, pp. 267-8.
7. Kuppusami Sastri, ed., *Srimad Valmiki Ramayanam*, Madras, Vijaya Press, 1958, Kishkinda Kaanda, Sarga 41, verse 15, p. 127.
8. Archaeological Survey of India, *South Indian Inscriptions* [hereafter *SII*], 26 vols., Madras & New Delhi: Archaeological Survey of India, 1890-1990, vol. I, no. 33, p. 29.
9. Gopinatha Rao, *Travancore Archaeological Series* [hereafter *TAS*], 9 vols., Madras & Trivandrum: Travancore State, 1910-41, vol. I, no. 54, verse 32.
10. *SII*, vol. II, no. 74, pp. 365-70; Udayendiram Plates, line 37, p. 367.
11. *SII*, vol. XII, no. 120, p. 58.
12. R. Rajalakshmi, *Tamil Polity*, Madurai: Ennes Publications, 1983, p. 191.
13. *SII*, vol. II, no. 74, pp. 365-70, Udayendiram Plates, line 62, p. 368.
14. P.S. Sundaram, trans. & ed., *The Kural*, New Delhi: Munshiram, 1990, chapter 2, verse, 8.
15. Ibid., chapter 2, verse 3.
16. Venkatasamy Nattar, ed., *Manimekalai*, Tirunelveli: South India Saiva Siddhanta Pathippu Kazhagam, 1992, XVII, verse 62, XI, verse 107, *Manimekalai*, cantos XXVIII, 1: 80 and XXIX 11:3: 25.
17. Venkatasamy Nattar, ed., *Silapathikaaram*, Tirunelveli: South India Saiva Siddhanta Pathippu Kazhagam, 1992, III, verse 37, see Avvai Duraisamy Pillai, ed., *Pathirrupathu*, 2nd edn., Tirunelveli: South India Saiva Siddhanta Pathippu Kazhagam, 1955, see Pathikam II.
18. U.V. Swaminatha Aiyar, ed., *Perungathai*, Madras, 1924, canto Unjai XXXVI, 235; see U.V. Swaminatha Aiyar, ed., *Puranaanuru*, Madras, 1971, song 9; see Venkatasamy Nattar, *Silapathikaaram*, XI, verse 18-20.
19. David Shulman, 'The Tamil Flood Myths and the Cankam Legend', in

Alan Dundes, ed., *The Flood Myth*, Berkeley: California University Press, 1988, pp. 312-14.

20. A. Appadorai, *Economic Conditions in Southern India (AD 1000-1500)*, 2 vols., Madras: University of Madras, 1936, p. 748.
21. K.A. Nilakanta Sastri, *The Colas*, 2nd edn., Madras: University of Madras, 1955, pp. 584-7.

CHAPTER 2

Waterscapes: The Rainfall in Tamil Country Ninth-Nineteenth Centuries

Water is one of the most dependable and continuing resources made by nature through rainfall. Thiruvalluvar in his *Thirukkural*, the Tamil work of the pre-Christian era, records that the world cannot exist without water. He says that every food of man is the gift of water and water itself is a food. He also says 'water from rains and springs, a mountain here and water thence; these make a land with fortress and defence'.[1] He devotes a chapter to rain ('Van Sirappu'), thereby giving it significant importance, next only to God, to whom he devoted the first chapter ('Kadavul Vazhthu'). We are well aware that rain and *aaru* (river) in Tamil country are looked upon as Gods because they are the sources of all fertility. It is mentioned in the odes of *Paripaadal* of the Sangam Tamil literature that the clouds absorbed water from the ocean and poured down to deluge earth. In the Sangam anthology of the Tamil text of *Purananuru* a reference is made to the importance of *kulam* (artificial tank) irrigation thus: 'Verily he who had turned the bent (low) land into a reservoir to arrest the flow of the running water is one who has established a name in the world'. Further, the construction of *karais* (bunds) for the storage of water is mentioned. It describes *karais* in the form of the three-fourths of the full moon.[2]

In the ancient period, Tamils were using natural resources available to them in a very limited way. They began to use them more in the middle ages and the infinite resources remained no longer infinite but became finite at that time. Man's intervention in nature was the chief cause of many natural disasters and he did not realize that. Rulers of Tamil country came to be called *kaaduvettis* (clearer of the forests) and we find that they cleared large and vast tracts of forests for human settlements and cultivation, which had been recorded in AD 808.[3] The vast tracts of forest reclaiming continued during the rule of the Vijayanagara rulers in the sixteenth century. People who reclaimed forest land had to pay 20 *kalams* of paddy on one *veli* of land including taxes such as *arasaperru* and *ilakkai*, further 8 *panams* had to be paid including taxes such as *kanikkai, sammadam* on each *veli* of land.[4] Thus, sizeable forest areas were reduced. This perhaps resulted in the absence of rains and resulted in famines owing to deforestation.

Agriculture and human settlements in Tamil country had considerably expanded during medieval times. Various *aarus* (rivers) such as Thamiraparani, Vaigai, Kaveri, Pennar, Palar, etc., were flowing and their immediate core areas emerged as the main centres of establishment and expansion of the various political powers such as the Pallavas, Pandyas, Cholas and Samburvarayas. The periphery zones too, later functioned as the backbone of agriculture. Irrigation in Tamil country, where the rainfall was scanty or occurred in the wrong season was of two types: natural and artificial. Natural irrigation was obtained through *aarus*, *yeris* (lakes), springs and rainfall while the artificial was by human effort in the form of excavated *kulams*, lakes and wells. In this period of expanding agriculture, we find harnessing of water on a large scale and an attempt is made here to examine the role of the rulers and the people.

THE DAM ACROSS KAVERI RIVER IN TAMIL COUNTRY IN THE THIRD CENTURY

Aarus have played a vital role in Tamil civilization. People cultivated the soil in the river valleys and derived sustenance from them. Agriculture played an important role in the economy. In the history of irrigation and flood control in the Tamil region, the control of the waters of the Kaveri river has been dated from the third century AD. The flow of water in the Kaveri had been blocked by throwing a *karai* upon the *aaru.* Mention of this construction of a monumental *karai* or barrage (18 ft high, up to 16 ft thick, 1,000 ft long) called the *Anaicut* (dam) on the banks of Kaveri river finds mention in the Tamil literary source of *Cholamandala Sathakam.*[5] It is also corroborated by an inscription at Tiruvaduturai.[6] The work was completed with the labour of 12,000 slaves who were brought after the conquest of Sri Lanka by the Cholas.[7] *Silapathikaram* the Tamil epic speaks of *anaikattu* (dam) with a door way and irrigated with the waters of River Kaveri. The text runs as follows:

> Finding her (Kaveri) movement arrested by the *anaikattu* she noisily leaps beyond it in the sportive mood natural to her finest freshes. No sound other than this can be heard. We can hear neither the sound of the bucket nor of water-lifts; neither the usually loud *picottah* nor the palm-leaf basket used in irrigation.[8]

An embankment which measured 80 miles along the Kaveri to divert water into irrigational canal systems 17 miles below modern Tiruchirapalli was in use.[9] Except this *anaikattu* on Kaveri river in Tamil country, we do not hear about the existence of any other *anaikattu* in that period.

The dam on River Kaveri had breached in AD 1596. Records mention that special funds were locally mobilized through a tax called *kallanai* and it was collected for the maintenance of the *anaikattu.*[10] The tax was applicable at Tiruvenkadu in Sirkaazhi area, Kunur in Kumbakonam area, Kuntalur in Nannilam area and Thanjavur region which shows that it was imposed on all the areas of the Kaveri delta. When this *anaikattu* breached, it was repaired with this special local fund.[11] We also find that this fund was called 'Kaviri karai thevai' (Kaveri riverbank fund).[12] Further, labour force was also demanded for repairing the embankments of River Kaveri. A tenant by the name of Keralantaka Vilupariayan had at first refused to send men on his behalf, but was compelled to send the necessary men for undertaking the repairing of the Kaveri river-bank.[13] Thus we find that many village assemblies during the early modern age had taken particular care in the construction of major *anaikattus* to store and supply water for irrigation purposes. These village assemblies chiefly felt the need for construction of new *anaikattus* to mitigate the woes of the people in the locality.

THE NEW DAMS BUILT IN THE SEVENTEENTH CENTURY

Portuguese missionary accounts throw light on the construction of an *anaikattu* in AD 1679 at Aanaikarai Palayam in Tamil country. The work was completed within three months and it was done for the development of agriculture in the area.[14] Further, an *anaikattu* at Suralur in Coimbatore region has been mentioned but no other details are available.[15] It is certain that topographical features were identified and noted for such *anaikattu* building endeavours as it could be easily constructed across *aarus* when they pass through small hillocks. However, very minute details and other salient features of *anaikattu* such as the maximum water level, dead storage level, length of the spillway *anaikattu*, total length of masonry *anaikattu*, top width and base width are not known from these records.

CONSTRUCTING A DAM ACROSS KAVERI RIVER IN AD 1701 BY THE KING OF MYSORE

Rights to the use of the waters of the Kaveri river had been reported by Pierre Martin (AD 1665-1716) the Jesuit in his letter of December 1703. He said that the king of Mysore blocked the waters flowing to Tiruchirapalli and Thanjavur by building a new *anaikattu* and he diverted the flow of water into his territory. In the first months of the year AD 1701 the strong southwest wind, the forerunner of the rains on the Malabar Mountains had begun to blow earlier than usual. The people of Tiruchirapalli and Thanjvaur, therefore, expected to get the Kaveri river in flood earlier than in previous years. Many days passed in expectation but the flood in the river did not occur. The reason of this delay was that the King of Mysore diverted the flow of water from the usual channel, to irrigate first large tracts of land in his kingdom. The ruler blocked the bed of the Kaveri by building an enormous dam across the entire breadth of it. He did not seem to care whether or not his plans would completely ruin his two neighbouring kingdoms in Thanjavur and Madurai. The water normally reached only by the end of July, and also dwindled by the middle of September. With their eyes open to the interests of their kingdoms, Mangammal (AD 1689-1704), the Queen of Madurai and Shahji II (1684-1711), the King of Thanjavur forgot their recent enmity and they combined themselves against the common enemy. They made fitting preparations to bring the king of Mysore by force of arms to break up the dam. However, before they were ready to move forward, the Kaveri river itself took upon to avenge the interest of the King of Mysore. The dam had burst and the water stored had swelled into great heights and flew into the territory of Tiruchirapalli.[16]

CONSTRUCTING A ANAIKATTU ACROSS KOLLIDAM RIVER IN 1836

An *anaikattu* was constructed across the Kollidam (Coleroon) river in 1836 under the auspices of Sir Arthur Cotton, the English Company official. Since the third century, the masonry barrage of the *anaikattu* had dammed the Kaveri river near its mouth at the Bay of Bengal at a great width. Therefore, in the nineteenth century, this *anaikattu* and the adjoining *kaals* (canals) had silted up, causing the water to be diverted from the main branch of the Kaveri river to the Kollidam. The latter was a substantial branch of the *aaru* that passed through the town of Chidambaram. In 1836, Sir Arthur Cotton undertook the construction of a new *anaikattu* across the Kollidam drawing lessons from the old *anaikattu*, and he adopted the technique of constructing masonry work over a foundation laid in the weak alluvial bed of the *aaru*. The *anaikattu* collapsed following in the year, and it was repaired. Yet it remained weak until it was completely reconstructed at the end of the nineteenth century.[17]

EXCAVATING YERIS (LAKES) IN THE MEDIEVAL PERIOD

In the absence of *aarus* and *sittarus* (tributaries) construction of *yeris* had been the only recourse for the development of agriculture in Tamil country. Several large *yeris* were constructed at different localities for storing rainwater in the medieval period. Vairamegha-tataka of Uttiramerur, the Chola-varishi of Sholingur, the Kaliyan yeri near Anaimalai in Madurai, the Periya yeri of Bahur near Pondicherry and the Rajendra Cholap-Periya yeri at Punganur may be mentioned.[18] Many individuals played a key role in this regard.

A rock inscription says that Parantaka Palli Velan had excavated a *yeri*; this rock inscription is in the village of Ramanathapuram in Dindigul region.[19] As the work was not finished during his lifetime, his son Pullan Nakkan undertook steps to complete it and it was called 'Pullan yeri'. He is acclaimed as the builder of many *yeris*.[20] A record of Anaimali in Madurai region registers the improvement of a *yeri* of Narasingamanagalam by a certain Arunuti Kailyan alias the Urudaiyan of Maruthur in Puramkarampainadu.[21] He is mentioned as having raised the banks of the *yeri*, besides lengthening and deepening them to irrigate the fields of the village and it was renamed as 'Kaliyan yeri'. In AD 1308, an individual called Kannan had built the Umayandi yeri at Pulavayal in Pudukottai region.[22]

SETTING UP A COMMITTEE FOR THE MAINTENANCE OF LAKES

Maintenance of *yeris* had been undertaken by local authorities in medieval times. The primary care of *sabha* (village assembly) was to get the silt removed on time from the *yeris* under their control, every year before the rains set in, to secure the proper depth needed to store the full supply of water for the coming year. The creation of a 'yeri vaariyam' (tank committee or board) in the village shows the keen interest the villagers took in the maintenance of the *yeris*.[23]

In Uthiramerur, the sabha had constituted the 'yeri vaariyam' and they spent a certain sum from the capital towards keeping the *kulams* in good condition.[24] Records refer to the election of members for the 'yeri vaariyam' of Uthiramerur.[25] Their main function was to look after the cultivation of lands granted for the maintenance of tanks (*yeri patti*). The income from this source was meant for the expenses incurred for the periodical removal of silt in *kulams* and also repairs and restoration. Another record mentions a gift of gold and stipulated that this was to serve as a capital for paying the ferry men to dig up silt from the bed and deposit it on the *karai* of the *peryeri* (big tank) and the excess produce that accrues out of the additional supply of water should be utilized for feeding the Brahmins.[26]

It is gleaned that special endowments were created in relation to some *yeris* to safeguard this important work from the neglect or the penury of village authorities. The annual renovations of the *yeris* were provided for by special local funds through collection of *yeri vari.* A cess called the *yeri ayam* was imposed on the peasants and collected in the Bahur and Tirubhuvanai regions of Pondicherry and the amount was used for the maintenance of *yeris.*[27] The sabha of Nemili in South Arcot region in AD 1010 had also set apart a certain amount for the maintenance of a local yeri.[28] A record also mentions that the expenses for the *yeri* maintenance was met out of *irai kaval,* the local tax collected.[29] Separate lands were also provided in each village under the name *vettaparru* (tenure for the *vetti*) and it was used towards the payment of the men who renovated the *yeris.*[30]

The village assembly of Vayiramega Chaturvedimangalam executed an agreement with two brothers from a neighbouring village because water did not reach the *kaal* flowing to the *yeri* in the village the brothers lived in. The village assembly, therefore, agreed to build a connecting *kaal* and allow the run-off water from their own *yeri* to travel through the *kaal* to the *yeri* of the neighbouring village. It was decided to remove the silt from the yeri and provide the labour.[31] Another instance is mentioned where a local man called Ariayan agreed to undertake repair works when there was a breach in the *yeri* at Thirukkanchi during heavy rains and storm in AD 1110. In the course of repair works, he even constructed a stone *karpadi* (revetment) on the banks.[32] It may, therefore, be said that the maintenance of *yeris* had been mainly the responsibility of the local people and the landholders.

THE ROLE OF VILLAGE HEADMEN AND LOCAL OFFICERS IN IRRIGATION

The *kilavan* (headman) of Sattur had renovated the old *karais* of Nerimadai yeri, Arasankal Cenkulam and Maran yeri. He dismantled the old *karai* of Palli kulam and built a new one with a *mathagu* (sluice; *mathagu* is a colloquial term for sluice while *thumbu* is the medieval term for sluice). His renovation of the Perungulam karai is mentioned in the Chinnakollaipatti area. Here the mathugu is mentioned as having been paved with stone and connecting the other two *vaykaals* (channels), *Ponnan madai* and *Punkunri madai* with Perungulam. The record ends by saying that all these works were executed by him in the region of Irunjolainadu. Individuals of different localities had evinced keen interest in the construction and renovation works of *yeris* and *kulams* in the Pandya country during the medieval period. We find detailed reference especially to the numerous *kulam* repairs executed by the *kilavan* of Errukankudi.[33] In one record of AD 827, the *kilavan* had dismantled an old *karai* on Karumkulam in Sattur and reconstructed it with a stone base. He renamed it as Kilavan yeri. In another record of 829, it is mentioned that he had excavated a *kulam* at Thenveliyankudi and called it Kilavan yeri. At Maranaur, he dismantled and rebuilt an old *karai* of Perungulam renamed it as Vaazhai kulam. He had also cut vaykkals with troughs for the flow of water at Iluppaikudi. At the same place, he connected two *karais* called *Kan madai* and *Paththan madai* in Thirumal yeri. He repaired a collapsed stone mathugu and constructed it with the help of string-line.[34] Similarly, Muri-Arulakki, a *kilavan* in the region of Anmanadu had built the *karai* and *vaykkal* of the *yeri* in Parankusapputtur. In AD 914, the *sabha* of a village near Srivilliputtur had decided to rename the *karai* and the *vaykkal* as Arulakkip-perum'atai and Arulakkip-perumkal in honour of the valuable services rendered by the headman to the people.[35]

In AD 857, a *kacham* (resolution) was passed by the Brahmin sabha of Ariksesarinallur regarding the operation of a *mathagu* under its jurisdiction. It mentions that the supply of water through the *kaal* called Srikanta vaykaal had to be done as per the decision.[36] Similarly, in AD 960, an undertaking was made by the *mahasabha* of Srivilliputtur to maintain the reservoir attached to the land purchased and endowed by a certain Satru Bhayankara Muvendavelan alias Velan Paranjothi.[37] It was specifically mentioned that this reservoir would be repaired if any damages occurred in the future by the village assembly.

Each agrarian unit at the village level had a network of *vaykaals* to supply water in each field, leading from the *ur kulam* (village tank) or the channel of the locality (*nattu perumkal*) to *yeris* and *kulams* which were either river-fed or rain-fed. Vasudeva *yeri* and certain Arrakai kulam and Nakkan yeri kulam existed in Nanguneri region of the Pandya country.[38] A *yeri* called Kon yeri was the chief irrigational source of Vaamadevamangalam, in Madurai region.[39] The *nattars* (officials of the nadu, territorial division) were involved in the adjudication of water rights because construction of larger *kaals* that cut through many villages and a number of nadus had been there.

Aandaar Arasukkadiyar excavated two *kaals* near Siddhalinga-madam after a joint meeting of the *sabha* members and the assembly of the Periyanattar was held. The land boundaries of these *kaals* in relation to surrounding paddy fields had been determined at that time, without lowering the water level in *kaals* going to a neighbouring village.[40] Besides creating new *kaals*, the reorganization of the existing local irrigational methods (like criss-cross patterns of feeder and drainage *kaals*) helped in bringing fallow land under cultivation to support a large immigrant population.

BUILDING THE CANALS

Long *kaals* or *vaykkals* were constructed and used for irrigation to carry water from its origin in a *aaru*, *yeri* or *kulam* through villages and often between villages.[41] A *kaal* was built for irrigation purposes from the Vaigai river. Construction of a *mathagu* is mentioned and the area irrigated by the new *kaal* reached Paruttikuttinadu (near Aruppukottai).[42] There is a reference in the same record to the closing of a *kaal* which had probably fallen into disuse. Manabaavan Muvenda Velan of Kappalur excavated another *kaal* called Thiyagachinyaperam from the Vaigai river to irrigate the lands in the region.[43]

Similar evidences mention that Siruthavur Udaiyan Velan Viranarayanan had constructed Killiyur kaal from the Kaveri river. He donated the irrigated lands to the temple at Thiruverumbur.[44] Sundara Pandya Kandiyadevan had dug a *kaal* and built a *mathagu* for carrying water of the South Pennar River for irrigation to the *devadana* (temple lands) in Nemmali village.[45] A *kaal* at *Melattamayamin* in Pudukottai area had breached in AD 1286.[46] It was repaired earlier. It again breached in AD 1301 and the needed repairs were made immediately.[47]

THE CONSTRUCTION OF THE WATER CHANNELS

Vaykkals (channels) were constructed in medieval times and in Thiruvayppadi village of Thanjavur region fallow lands were sold in order to construct and raise a *kulam karai* and also to dig a *vaykkal* for the irrigation of the cultivable land.[48] Owing to the silting of an irrigation *vaykkal*, the sabha of Nerkunram had to divert waters from a spring in a neighbouring village after giving compensation to it.[49]

Olukkai was a branch channel from a main *kaal* that watered specific fields or small plots of lands. A *vaykkal* called Paramesvara vaykkal near Vijayanarayanam was passing through the Brahamadeya village.[50] Another main *vaykkal* of irrigation in the village also finds mention in a different record.[51] Three main *vaykkals* called Alaikkal, Marutakuna vaykkal and Thiruvenkata vaykkal carried water in Suchindram area to the devadana of Nripasekhara valanallur in Pandya country.[52] The Uyyakondan irrigational *vaykkal* across the Kaveri river was in existence in Thiruchirapalli region. It is a remarkable piece of native engineering work which is still in a good condition, supplying water to the town of Tiruchirapalli. It is the massive head *mathagu* of the Uyyakondan vaykkal, a branch of River Kaveri which is made of a solid granite block and was erected in AD 1205-6 by Kulothunga Chola III. Another example is the head *mathagu* of the *peru vaykaal* (big channel) on the banks of Kaveri river near Musiri and this was spanned by a bridge across the vaykkal. On one of the side walls, Chola Raja Raja III and the eighth year of his reign (AD 1219) is mentioned and the head *mathagu* was built of stone at Musiri.[53] New *kaals* were dug at Tiruvaalangadu during AD 1012-14 that fed the minor irrigation *vaykkals*.[54] Saluva Thirumalai Devaraya had constructed two *kaals* and he created the biggest network of *vaykkals* to irrigate lands in the Karaikal and Tirumalarayanpattinam areas.[55]

In Pandya country, during the rule of Jatavraman (AD 645-70) construction of *mathagu* and excavation of *vaykaal* from Vaigai river had been carried out. Aaru irrigation had continued also during the time of Srimara Sri Vallabha (AD 811-57) and we find in records that many old *vaykkals* were either remodelled or renovated and several new waterworks had been added to dry areas.[56] Pallavarayan, a local official, had built a *vaykaal* in AD 1221 at Alagarkoil near Madurai. Jaya Simḥa, another official, had cut a *kulam* (here *kulam* denotes a pond) and *vaykaal* in AD 1286 at Dalavanur.[57] In the Kanyakumari region, Tennavan had cut a *vaykaal* through the lands purchased by him in Suchindram.[58]

BUILDING WATER TANKS, TANK BUNDS AND SLUICES

Water from the *kulam* (here it denoted a tank) was taken to fields through mathugu. New *vaykaals* were constructed near an existing *kulam* at Thirukacchur by an individual. A fresh *mathugu* was erected and funds were made available from the treasury of the temple so that the devadana could be irrigated since the *kulam* originally belonged to the people of Sengunram and their consent was also obtained before the extension of works were undertaken. The water in the *kulam* was also agreed to be distributed between the villagers and the temple in the ratio of their landholdings.[59] The existence of Perungulam in AD 808 at Perumpalanji in Nanguneri region is known and its *kumili* (sluice pit) was called *velvi kumili*.[60]

In numerous places, *kuttai* (small ponds) were also constructed for irrigation and the *mathugu* became part of the tank-fed system. These *mathugus* consisted of two granite pillars varied in height, according to the depth of the reservoirs. The mouth of the *mathugu* could be easily opened by lifting the rod. The *kumili* was another system for the regulation of the flow of water from the reservoir used at required centres. In the case of *kumili*, water from the reservoir flowed into the *mathugu* from above.

In AD 819, Etti Saathan, the *kilavan* of Iluppakkati was the renovator of various *kulam karais* and *mathugus* in the Chinnakollaipatti area.[61] He had rebuilt in granite the collapsed *karai* of Arasaiyur Kal in Sattur. In the same place, he built two more *karais* in Peru-mkulam called Ponnan madai and Punkunri madai. He also rebuilt sennir madai (rainwater karai) in bricks at Puthukulam. Another individual called Alagaiya Pallavarayar had repaired the Illupai madai and Sappai madai at Kunnathur in AD 1222 near Madurai. There was a stone mathugu called Paala madai at Nilaiyur in the same region.[62] Nambi Puduvan had built a mathugu in AD 1251 at Thirumangalakurichi.[63] Further, Sankaran Vasudevan had repaired the damaged *mathugus* because of the heavy floods in AD 1303 at Vijayamangalam in the Tirunelveli region.[64]

The term *kulam* and its various types frequently appear in medieval Tamil inscriptions.[65] *Pulattir kulam* (tank in a cultivation field), *kalani kulam* (tank in a paddy field), *uruni kulam* (common tank for drinking water), *uruni naduvupatta kulam* (tank situated in the centre of the village), *thirumanjana kulam* (sacred tank for purification), and *parai kulam* (water pool for the paraiyar lower castes) are important to note.[66]

Choosing the most suitable points and places where *kulam karai* and banks should be raised for storing the maximum quantity of water besides fixing the position and depth of the *mathugus* had been an art well known and practised in medieval Tamil country. The inscription mentions that chiselled stone was used in Ramanathapuram for the construction of the *mathugu* in Pullan yeri.[67] The kilavan of Illupaikudi had set the stone straight by the string-line method.[68] The use of chiselled blocks of stone instead of rubble and a string-line to set the stone were the two notable improvements in irrigation technology. Inscriptions do mention the removal of rocks while constructing *kaals* in this period.[69] *Thumpu* (a single valve system) was operated without any device for mechanical advantage. In this system, water from the reservoir flowed into the sluice from the side. We also find water lift that had been worked out by bulls. There had been a ban in some areas to the use of baskets for lifting water and it has been stated clearly in the inscription.[70]

DIGGING WELLS

Kinarus (wells) were a source of irrigation in medieval south India. The construction of wells had been much less prevalent even in dry areas that were far away from *aarus*. However, we find use of wells for irrigation in individual plots of land. *Naaladi kinaru* and *irandadi kinaru* were dug. These suggest respectively that they were with a depth of four feet and also with a depth of two feet. The Dalavaypuram copper plates refer to the irrigation of fields with water from wells attached to them.[71]

If an individual desired to dig a well for irrigating his land, he had to get a formal approval from the village authorities by paying a small amount known as *ulliyak-kuli* (well-diggers wages). This is mentioned in the Kasakudi plates of Nandivarman II.[72] Digging such large wells for the purpose of irrigation was strictly under the administrative control and this system had continued by levying a small sum known as *ulliyak-kuli*.[73]

In his *Yerelupathu*, Kambar, the medieval Tamil poet, laid emphasis on the importance of *yettam* (piccottah). He said that there will be no rain if the kings failed to rule with justice. There will be famine everywhere if people failed to undertake their own duty faithfully. Whatever may be the adverse conditions there will be no hunger, if the peasants did their job properly, by watering their crops and plants by drawing water from the wells through piccottah.[74]

Kinarus were attached to fields mainly to supplement irrigation from the *yeris* and *kulams*. We find mention of *thotti* (cistern) too. *Kinarus* became an important source of artificial irrigation in some areas under the rule of the nayak rulers. Masonry *kinarus* or *kinarus* without use of bricks were abundant. Generally, these *kinarus* were used for drawing the drinking water. Manuel da Costa, the Portuguese missionary, had dug a *kinaru* at Kanavanakkari at his own expense in AD 1677. He also made arrangements for cleaning another *kinaru* in order to get good drinking water for the people of the place.[75] Some nayak rulers focused their attention on providing irrigational facilities and invested funds in the creation of minor irrigational systems which brought about agricultural expansion in their kingdoms.[76]

THE WATER RIGHTS AND THE DISPUTES

Mention of the water rights attached to particular plots of land had been a common feature since landholdings changed hands frequently by sale or gift. Expressions such as liberty to dig *kinarus*, *kaals* and *vaykkals* in accordance with watering requirements, not to waste *sennir* (rainwater) but to store it for irrigation are found mentioned in various stone records. The Thiruvalangadu copper plates confine this restriction to persons other than the grantees.[77] The right to raise the *karai* of the *kulam* in a village to its maximum height and to store in it the maximum quantity of water had been only vested with the *sabha*.[78] Puvanan Paraiyan who purchased a barren plot of land in the periphery of Brahamadeya called Ilango-kudi in Mullinadu had to excavate a small *kulam* for irrigation of the land.[79] Several instances of the purchase of barren lands by individuals for endowment to temples have been reported in the records only after providing necessary irrigation facilities.

WATER MANAGEMENT UNDER THE NAYAKS AND SETHUPATHI RULERS, SIXTEENTH-SEVENTEENTH CENTURIES

The nayak rulers focused on providing irrigational facilities and invested funds in the creation of minor irrigational systems, which brought about agricultural expansion.[80] Nayinappa nayak made lands cultivable by digging wells at various places in South Arcot region,[81] and at Devikapuram in North Arcot area.[82] Ramappa nayak, Tirumalai nayak, Ellappa nayak, Bhaiyappa nayak and Sriranga nayak improved certain tracts of land in Tirukkadaiyur,[83] Tirupati,[84] Chandragiri[85] and Padaividu[86] by excavating new *kulams* and *vaykkals*.

Whoever financed or executed the construction of such irrigation works, received a small share of the produce for his pains, while the major share of the enhanced produce went to the cultivators.[87] The right of the former, called *dasavanda*, was personal, heritable and transferable.[88] Stone records mention that a piece of rent-free land was granted for building or repairing a kulam on condition of paying one-tenth of the produce.[89] The nayak's motivation for constructing irrigation works came partly from these material gains, and partly from the spiritual merit this was supposed to bring, since such an act was among the seven *satantaka* acts which brought *punya* and ensured one's happiness in the next world.

For maintaining existing irrigational works, nayaks raised funds through various sources. One source was state funds and incomes of fish lease from *kulams*. Thus, Sevappa nayak allocated the income from the lease of fishery in the *kulam* at Kodungalur for deepening the *kulam* in Nedungunram locality.[90] Pottu nayak gave the money from fish lease of the Senalur kulam in South Arcot area to be spent on the *kulam* itself.[91] Damal Kama nayak ordered that money realized by the sale of fish from Sripurushamangalam kulam in Cheyyar locality should be spent on repairs of the *kulam*.[92]

The nayaks also invested their personal funds in improving *kulam* irrigation. There was a system of the donor's share at the temples called *vittavan vilukaadu*. In this, one-fourth of the sacred foodstuffs offered at the temple were given to the person investing in the improvement of the irrigational resources of the temple.[93] A Tamil inscription speaks of a deed called *silasasanam* with this provision made by the *sthanattar* (temple superintendent) of the temple in favour of Mallappa nayak of Nedungunram. This assignment was valid for all generations, as long as the sun and moon existed.[94] This probably resulted in the regular trade in *prasaatham*; in one instance, such a sale reached 4,600 *panams*.[95]

The rate of *nirkuli* (irrigation tax) levied on the lands in Kodangi, Nellikuppam, Timmanankuppam, Vengalakuppam and Palaveri was fixed by Lingamayya nayak in the Chingleput area.[96] Nirkuli was also levied on *maruthams*, dry lands, garden lands and groves in the Tiruppunduritti area by Achutappa nayak of Thanjavur.[97] References to the levy and collection of *nirkuli* by nayaks in Chandragiri and Chittoor,[98] and removal of water cess by nayaks in Sittamur and Gingee by royal proclamation,[99] are indicative of the control exercised by nayaks. Some of these irrigation taxes were gifted to temples.[100] An epigraph suggests that an officer designated as *jala-sutrada*, by the name of Tiruvenkata Chirukkan supervised the digging and maintenance of *kaals* and *vaykkals* in the department of *varigripakarana* (water works).[101] This challenges the traditional views of N. Venkataramanaya and Burton Stein that there was no department of irrigation or public works in the Vijayanagara domain.[102] Irrigation was, however, a local subject included in the nayak administration; the centre in Vijayanagara was not much concerned with it.

Where more than one village depended on the same source of irrigation, the management of the irrigational system was collective. This probably led to water disputes. A dispute arose between temple authorities and the residents of the locality when one irrigational channel was dug in Tirupati, which was settled by an officer named Yagnarasar appointed by a nayak. This officer inspected the site and found the objections raised to be legitimate.[103] Another irrigation dispute between the Srirangam and Jambukeswaram temples was decided by the officer Mallar Ayyangar appointed by a nayak, who directed that the irrigation *kulam* in Sembiyanallur should be used by the Srirangam temple.[104] Similar evidence also describes an irrigation agreement made between the trustees and treasurers of the Siva and Vishnu temples of Perumbakkam, which permitted the former to dig a *vaykkal* within the limits of Perumbakkam to carry water to the *kulam* of Tiruvamathur in exchange for 300 *kuzhi* (81 sq. ft) of *maruthams* as compensation in Vedampattu village. This deed was drawn in the presence of Bommu Reddi, an officer appointed by Achutappa nayak who settled the dispute.[105] In another instance, an order issued by a nayak gave the benefit of irrigational *kaal* flowing through Akkalimangalam and Pudupalayam when the people of Iraiyur village declined to use it.[106]

Water from River Vaigai was regarded as essential for irrigation. Hence, Kilavan Sethupathi (AD 1622-33), the ruler, excavated a *kulam* in AD 1628 at Mudalur village near Kulathur and he brought water from the north of Vaigai river through a *kaal* and water was diverted into the areas of Paramakudi and Muthugulathur. This *kaal* was named after him and it was called Kuthan Canal.[107] Further, when the water from the Vaigai river was blocked from flowing into Abiramam yeri, the ruler Muthu Ramalinga Vijaya Raghunatha Sethupathi (AD 1760-95) had marched with his army and cleared the blockages which subsequently helped in the irrigation of the area.[108]

THE SCIENCE OF READING THE RAIN BY JOHANN ERNEST GEISTER IN MADRAS, AD 1732-7

European travellers in the Tamil region observed clouds, tides, gales, winds, breeze and all other meteorological events. Thomas Bowrey the English traveller who visited Madras in AD 1669, gave details of the monsoon and the climate. He mentioned that in the months of May and June, people enjoyed the benefit of sea breeze. There had been fresh gales which were somewhat sulphurous, and this he attributed to the heat of the sun.[109] An English Company official at Madras reported on 22 August 1676 that the tide rose and fell from 2.5 ft to 3 ft at the springs.[110] Joseph Collet the English Governor of Madras wrote to his brother in AD 1718 and described that the sea breeze of the port of Madras had given him renewed vigour and that he was in perfect health.[111]

Johann Ernest Geister, a German missionary attached to the Danish-Halle mission, sailed from England on 1 March 1732 and reached Madras on 26 July 1732.[112] He maintained a weather diary from 20 October 1732 to 20 July 1737.[113] Geister painstakingly and meticulously recorded and described the rainfall each day. He gave precise details about the timing and nature of rain with a lot of references to night rainfall. Geister used various words to differentiate the types, amounts and intensities of rainfall. Geister's diary commenced with a general summary of the weather followed by the systematic wind and rainfall records. Apart from the daily recordings and frequent observations of wind direction, he also noted cloud cover, precipitation and made occasional remarks on wind strength, thunderstorms and temperature. Each year, a section titled 'General Observations' in his diary contained an annual review of climate with comments on harvests and food prices. Geister stated that the weathervane was put up at some 30 paces from the beach between the houses in Madras.[114] According to Geister, the maximum number of rain days in the entire period occurred in June or July. The average rainfall frequencies were sharply lower in the south-west monsoon months in AD 1733-7 and in the north-east monsoon months in AD 1732-7.[115]

Some general comments made by Geister in his diary are noteworthy. He commented upon the comparative rainfall during AD 1732 and observed that at the beginning of October, the wind, which

TABLE 2.1: RAINFALL ESTIMATES IN MADRAS IN MILLIMETRE FOR 1733-6

Month	1733	1734	1735	1736
January	7.9	15.8	31.6	0.0
February	0.0	0.0	13.6	0.0
March	0.0	60.4	0.0	60.4
April	24.8	12.4	12.4	12.4
May	17.2	34.4	103.2	17.2
June	63.8	63.8	121.8	40.6
July	126.0	72.0	90.0	96.0
August	124.5	58.0	107.9	99.6
September	139.1	96.3	96.3	107.0
October	171.9	210.1	191.0	114.6
November	56.2	337.2	309.1	112.4
December	19.9	39.8	79.6	39.8

Source: R. Walsh, R. Glaser and S. Militzer, 'The Climate of Madras during the Eighteenth Century', in *International Journal of Climatology*, vol. 19, issue 9, July 1999, pp. 1025-47, see p. 1033.

until that time had been from the south and west, changed to a north-easterly direction. In the middle of October, there was rainfall, and it was particularly heavy for three to four days. On 17 October 1732, the wind changed again to southerly direction silently, and was accompanied with rainfall especially at night. The final months of 1732 saw very heavy and considerable rainfall, which was a sign of a good harvest.

Again, he mentioned the unusual rain of the south-west monsoon and the dryness of the north-east monsoon of 1733. He commented that in July, August and September of 1733, more rain fell than was normal at that time of the year. In contrast, in the proper rainy season, which occurred in October, November and December, very little rain fell in Madras and to the north and south of Madras, and prices rose considerably as a consequence. He also alluded to the somewhat higher rains of the north-east monsoons of 1734 and 1735, and said that the proper rainy season in 1734 was much stronger than the previous year, but was not complete, because the soil, which had become very dry earlier on, as well as the dry ponds or *kulams* and *aarus*, required a great deal of rainwater.

He wrote in AD 1735 that the rainfall in November and December were unusually not high, but sufficient for the fertility of the land and filling of ponds. He described the dryness of both the monsoon seasons in AD 1736 and said that in the whole year there was little rain; land wind remained very strong until the end of the summer and at the end of the year there was a great increase in prices. This aridity stretched from Madras 100 miles northwards and over 10 miles southwards, and quite a few days' journey inland.[116]

Geister had discussions with the residents of Madras in AD 1735, and recorded that not only in the early 1730s, but also for 20 years prior to AD 1735 the weather had been considerably dry in the Madras area with less rainfall than normal. At first, he was unwilling to accept this account, considering that it was merely an exaggeration of the old times and underestimation of the present. However, he was compelled to accept this as the truth when he received confirmation from different persons, who made reference to their own experience and who had no interest in strengthening or confirming such a view. Formerly, almost six whole months were rainy, and sometimes rain fell up to 20 days and nights without stopping, so that the inhabitants of the area would bid goodbye to each other before the rainy season, because they were unable to visit each other at the time of great water. It may be pointed out that the wind directions (land breeze at 8 hours and sea breeze at 17 hours) in the Tamil coast exerted an influence on the rainfall. In his study, Geister made no mention of these two timings.

RAINFALL IN PONDICHERRY IN THE EIGHTEENTH CENTURY

The diaries in Tamil written by Ananda Ranga Pillai, Rangappa Thiruvengadam Pillai and Muthu Vijaya Thiruvengadam Pillai provide details of rainfall in Pondicherry. It is easy to classify the period of the monsoon annually as excess rainfall, heavy rainfall, medium rainfall, less rainfall, no rain and floods from these sources. These Tamil sources furnish information on the rainy days and the nature and actual condition of rainfall. These are given in the Table 2.2, 2.3, 2.4.

TABLE 2.2: RAINFALL IN PONDICHERRY, 1761-2

Date	Details
1 January 1761	Storm
29-31 October 1761	Rain
1 November 1761	Cyclone
4-5 November 1762	Rain
23 July 1762	Rain and Cauvery in spate

Source: S. Jeyaseela Stephen, *Rangappa Thiruvengadam Pillai Naatkurippu, 1760-6*, II vols., Pondicherry: Pondicherry Institute of Linguistics and Culture, 2000; S. Jeyaseela Stephen, *The Diary of Rangappa Thiruvengadam Pillai, 1760-8: Translated from original Tamil with Notes*, Pondicherry: Institute for Indo-European Studies, 2001; S. Jeyaseela Stephen, *Arangappa Thiruvengadam Pillai Naatkurippu, from 13 June 1767 to 29 December 1769*, vol. III (in Tamil), Chennai: Sekhar Publishers, 2006.

TABLE 2.3: RAINFALL IN PONDICHERRY, 1794

Date	Details
5 October	Full day rain
6 October	Full day rain increased
23 October	Intense rain
24 October	Drizzling
28 October	Raining from morning
29 October	Heavy down pour
30 October	No sunshine, raining could not be described
31 October	Upparu over flew, Usteri lake over flew
17 November	Drizzling whole day
18 November	Rained
29 November	Cloudy, no sun
30 November	Rained
8 December	Heavy rain
9 December	Flood, Usteri lake full, old walls fell off
10 December	Rain stopped

Source: S. Jeyaseela Stephen, *The Diary of Muthu Vijaya Thiruvengadam Pillai, 1794-6*, Pondicherry: Institute for Indo-European Studies, 1999.

When we compare the rainfall data, we notice that rains started in Pondicherry on 5 October 1794. In AD 1795, it started on 31 October and so the rainy season was delayed by 25 days. Similarly,

TABLE 2.4: RAINFALL IN PONDICHERRY, 1795

Date	Details
31 October	In the afternoon starts raining
27 October	Drizzling
1 December	Rains and trees fell
5 December	Flood at Arcot, Kovalam, Sadras & Marakkanam
21 December	Heavy rain and storm
24 December	Heavy rain from midnight
29 December	Rained and drizzled
30 December	Rained

Source: S. Jeyaseela Stephen, *The Diary of Muthu Vijaya Thiruvengadam Pillai, 1794-6,* Pondicherry: Institute for Indo-European Studies, 1999.

rains stopped in Pondicherry on 10 December 1794. In 1795, it stopped on 30 December. Hence, the rainy season had been extended by five days more between the two years of AD 1794 and 1795.

USING THE RAIN GAUGE TO MEASURE THE RAINFALL AT MADRAS, 1803-25

The use of rain gauge was introduced in Europe in AD 1676 and it spread to many parts of the continent by 1690. Roxburgh in Madras had mentioned that he began to use the rain gauge in AD 1776, but said that the rain gauge was of indifferent quality, so he could not measure the rainfall with certainty.[117] In 1802, English East India Company used the rain gauge in Madras to measure the rainfall. It was a funnel of copper, 12 inches in diameter, and was inserted in a *chatty* (a vessel in Tamil). After the rain stopped, the water in the *chatty* was emptied into a copper cylinder 8 inches deep and 3 inches in diameter. As the diameter of the funnel was four times that of the cylinder, the area of the former was 16 times that of the cylinder, so that when the rain water was poured into the cylinder, every inch showed one-sixteenth of an inch of rain had fallen. Thus, a full cylinder or a measure denoted 0.5 inches of rain. A dipping stick, diameter to inches and tenths, was used to

TABLE 2.5: THE MEAN FALL OF RAIN IN INCHES AT MADRAS FOR THE YEARS BETWEEN 1803 AND 1819

Months	Including the whole fall during storms	Fall in storms reduced to the mean fall
January	0.608	0.608
February	0.127	0.127
March	5.384	5.384
April	0.384	0.384
May	1.419	0.121
June	0.646	0.746
July	3.303	3.303
August	3.552	3.552
September	4.824	4.824
October	11.294	11.294
November	14.803	14.803
December	8.618	6.048
Mean	50.124	46.348

Source: John Goldingham, 'Results of Meteorological Inquiries, Made at Madras', in *Transactions of the Royal Asiatic Society of Great Britain and Ireland*, vol. 3, no. 3, 1834, Appendix no. III, p. xxvii.

measure the portions of the cylinder. One inch in the cylinder, as stated before, denoted one-sixteenth or .0625 of an inch of rain and one-tenth of an inch deep in the cylinder denoted 1/160 or .00625 of an inch.[118] The Europeans, thus, verified the results with apparatus and scientific measuring tools of their own creation.

The mean fall of rain during the monsoon season, as this was termed, was about 33 or 34 inches. The fall in a year, as given in the above statement, was 50.214 inches, including the fall during the storms which had occurred some years, but only about 46.35 inches, if the fall was reduced to that of ordinary years. The statement alluded to shows the fall of rain for 13 years, while in the following statement some years are added, making altogether 18 years.[119]

If the fall during the storms were allowed, the yearly fall would be 46 inches, or about one-third of an inch less than that given earlier for 13 years.

TABLE 2.6: THE MEAN MONTHLY FALL OF RAIN AT MADRAS INCLUDING THE FALL DURING THE STORMS IN EIGHTEEN YEARS, 1803-25

Months	Inches	Months	Inches
January	0.739	July	2 .945
February	0.099	August	3.883
March	0.469	September	4.359
April	0.333	October	12.273
May	1.354	November	13.937
June	0.854	December	7.522
Mean Fall Annually			48.755

Source: John Goldingham, 'Results of Meteorological Inquiries, Made at Madras', in *Transactions of the Royal Asiatic Society of Great Britain and Ireland*, vol. 3, no. 3, 1834, Appendix no. III, p. xxvii.

The south wind commenced about the 2nd of March and blew along the shore, bringing with it a great degree of dampness, and having at the same time sultriness. This wind blew, according to the average standard before-mentioned, till the 29 April, when there were sometimes for a week or two land or southwest and westerly winds; and at other times south and south-easterly winds. The land wind set in about the 16 May, and continued blowing, generally with a great degree of heat, during some weeks, cooled at intervals, however by showers; it afterwards prevailed only at night, and in the early part of the forenoon, when during the remainder of the day its place was supplied by the southeast or sea breeze.[120] About a month or more before the monsoon and commencement of the rains, the wind was variable, with calms and a sultry and oppressive state of the atmosphere.

Goldingham gave the average times of the commencement of the monsoons but said that there were considerable variations in a stated period of years. In the interval under review, for example, the north-east monsoon and rains set in one year as early as 29 September, and another year as late as the beginning of November. The land wind also commenced at one time as early as the end of April, and kept off at another time till the beginning of June. Respecting the state of the atmosphere, the results of these meteorological observations showed it was not quite so clear and serene as was commonly supposed; the mean of 26 years had given during the sun's revolution only 180 clear days; the remainder of the year was made up of 64 days clear, hazy and cloudy in some part of the 24 hours; of 96 cloudy days, and 25 hazy; there were also 57 days on which rains fell during the year; 31 days when there was a dew; and only 18 with lightning. The statement, being the mean of the details of the weather in the diary for the foregoing period of 26 years, showed how this state of the atmosphere was divided among the different months. These findings were constructed by Goldingham from a great mass of detail.[121]

In 1856, Bayley suggested that the rain gauge used for measuring the rainfall in Madras by the English Company needed modification. He said that the imperial fluid ounce consisted of 1.733 or more exactly 1.7329625 cubic inches. Consequently, the fluid ounce measured necessarily represented the tenth part of the bulk of rain falling at the rate of 1 inch in any given period, on an area of 17.33 square inches. The proposed arrangement was scientific and its precise accuracy merely depended upon the truth of the workmanship and the care of the observer.[122] On Colonel Smith's recommendation, six rain gauges answering to Bayley's description were ordered, to be made up and tested at the English Company's Observatory. Major Worster, the Company's astronomer, referred to it for use.[123] The government then approved of the new rain gauge and sanctioned measure of 180, with the input suggested in the letter of the chief engineer.[124]

RAINFALL STATISTICS IN MADRAS, 1803-87

While missionaries had carried out climate studies for short intervals in the Tamil coast, the English East India Company began to engage in the study seriously with the use of instruments for long periods. The servants of the English Company used rain gauge, barometer, thermometer and hygrometer and their observations of weather and climate included rainfall, surface air, sea-surface temperature and atmosphere pressure. They carried out studies under two different systems: terrestrial and marine. Rainfall statistics was collected at Fort St. George. The Madras register for rainfall extends over 74 years from 1803-77, with the rainfall in Madras being measured in inches.

Hunter used the records of the rainfall at Madras from 1813-77 (a period extending over 64 years) to establish a cycle of rainfall which had a marked coincidence with a corresponding cycle of sunspots—the rainfall and sunspots attained a minimum in the eleventh, first, and second years, and a maximum in the fifth year.[125] Richard Strachey, the chief of the meteorological office in London in 1876, after reading Hunter's writings, opined that irrespective of its scientific interest, the conclusion adopted would, if sound, be of no little practical importance, as it supplied a means of indicating the probable recurrence of those seasons of excessive drought which produced such terrible results, and from one of which the Madras area was then suffering in an extreme degree.

Strachey, therefore, made a very careful analysis of the data. He first referred to the Madras observations and then Hunter's results. He stated that the Madras register extended over 64 years,

TABLE 2.7: RAINFALL IN MADRAS IN DAYS AND INCHES, 1827-76

Month	Number of Days	Rainfall in inches
January	3	1.0
February	1	0.3
March	1	0.4
April	1	0.6
May	3	2.2
June	10	2.1
July	14	3.8
August	14	4.4
September	11	4.7
October	14	10.8
November	14	13.7
December	9	5.1
Total	95	49.1

Source: Henry F. Blanford, *A Practical Guide to the Climates and Weathers of India, Ceylon and Burma, and the Storms of Indian Seas*, London: Macmillan, 1889, p. 329.

TABLE 2.8: DAILY RAINFALL OF GIVEN AMOUNT IN INCHES IN EACH MONTH OF THE YEAR DURING 1875-87 IN MADRAS

Month	Below ¼	¼ to ½	½ to 1	1 to 2	2 to 3	3 to 5	5 to 7 ½	7 ½ to 10	Above 10	Highest
January	21	3	4	2	–	–	–	–	–	1.3
February	1	–	–	–	1	–	–	–	–	2.5
March	7	2	3	1	–	–	–	–	–	1.2
April	2	1	2	1	–	–	–	–	–	1.7
May	21	4	3	6	–	1	1	–	1	13.0
June	101	20	9	8	–	–	–	–	–	2.0
July	129	20	21	10	–	–	–	–	–	1.9
August	123	30	19	11	4	–	–	–	–	2.8
September	104	16	17	7	6	2	–	–	–	4.0
October	98	29	21	24	12	9	1	–	–	5.8
November	73	34	18	34	17	12	6	1	–	8.2
December	67	13	16	13	6	4	1	–	–	5.1
Total	747	172	133	117	46	28	9	1	1	13.0

Source: Henry F. Blanford, *A Practical Guide to the Climates and Weathers of India, Ceylon and Burma, and the Storms of Indian Seas*, London: Macmillan, 1889, p. 262.

beginning with 1813, and the mean rainfall for the whole period was 48.5 inches. The deviations from the mean varied from 30.1 inches in deficit to 39.9 inches in excess. The arithmetical mean of these deviations (disregarding the signs) was 12.4 inches. The greatest difference between two consecutive years was 50.5 inches, and the average difference 15.8 inches. Hunter, in order to test the point which he proposed to investigate, divided the 64 years' observations into six cycles of 11 years, and calculated the arithmetical mean of the successive years of the whole series of cycles and gave the results in the table, the figures in which represented the differences of the several years' observations from the mean of the whole.[125]

In these calculations, the first year of the cycle of 11 was 1813, so that the average period of maximum sunspots would be about the third or fourth year of the cycle, and the period of minimum would be about the tenth or eleventh of the cycle. The table seemed to indicate a period of maximum between the third and the seventh years, and of minimum between the eighth and the second years. But to estimate the true weight of these results, Strachey looked a little deeper. Then, the only signification of the arithmetical mean of a series of observed quantities was that it was a quantity above and below of which there was an equal amount of deviation in the aggregate of individual observations. Further, conformity to a law of any sort in a series of observations in relation quantity had to be obviously tested by the extent of deviation of the observed quantities from the results that such law required. Consequently, in Hunter's study, the question whether or not the mean values shown could be accepted as showing a definite law of variation from year to year, the cycle had to be determined by examining the differences between those means and the individual observations on which they were based. Thus, it appeared that the mean difference of the individual observations from the calculated means shown differed but little from the mean difference of the individual observations from the arithmetical mean of the whole series.[126] In other words, the supposed law of variation obtained from the means of the six 11-year cycles hardly gave a closer approximation to the actual observations than was got by taking the simple arithmetical mean as the most probable value for any year.

In order to obtain a practical test of the probable physical reality of the cycle of 11 years, Strachey calculated a series of mean values corresponding to those given for a series of cycles of five, six, seven, eight, nine, ten, 12 and 14 years. He found that the mean differences between these means and the observed quantities, and therefore, corresponding to the mean differences shown were all within a very small fraction of one another and of the mean obtained from the 11-year cycle. In short, that one cycle was in this respect almost as good or as bad as another. So, if in any series of quantities, such as the rainfall observations at Madras, there was a law of periodicity, each observed quantity might be supposed to be compounded of a periodical and a non-periodical element. If they took the sum of a large number of cycles, each of which coincided with the cycle of periodicity, the non-periodical elements would tend to occur in equal amount in excess and deficit, and, thus, to be eliminated, and the means for the successive years of the cycle, or whatever the intervals be (which Strachey termed cyclical means), would tend to indicate the periodical elements for the successive intervals. At the same time, the differences of these cyclical means, from the several original quantities from which they were obtained, would approximate to the several non-periodical elements.[127] These differences Strachey called cyclical differences.

In proportion as the periodical elements were small or large in relation to the corresponding non-periodical elements, so the cyclical differences would be inversely less or more different from the differences between the individual observations and the mean of the whole of them. And if there were no periodicity, the cyclical mean would tend to disappear, and the two sets of differences would, in a sufficiently long series, be identical. Hence, it might be inferred that when the cyclical differences closely approximated in magnitude to the mean difference of the original observations from the arithmetical mean of all of them, the periodical elements in those observations might be correspondingly small. This applied manifestly to the whole of the cycles for which the differences were shown.

The mean rainfall on drainage basins of *aarus* in Tamil country had also been noticed. The tables in the official reports of the meteorological office for the period of 60 years from 1817-76 had been used to study the rainfall.

TABLE 2.9: THE MEAN RAINFALL ON DRAINAGE BASINS OF RIVERS, 1817-76

River	Rainfall in inches	River	Rainfall in inches
North Pennar	26	Vellar	40
Palar	36	Kaveri	44
South Pennar	33	Vaigai	32

Source: Henry F. Blanford, *A Practical Guide to the Climates and Weathers of India, Ceylon and Burma, and the Storms of Indian Seas*, London: Macmillan, 1889, pp. 284-5.

THE RAINFALL STATISTICS IN TIRUCHIRAPALLI, MADURAI, COIMBATORE, PALAYAMKOTTAI AND SHENKOTTAI FROM 1827-76

The English East India Company had set up facilities to measure the rainfall in inches at some important places like Tiruchirapalli, Madurai and Coimbatore. These statistics are also useful for the study. These details are found in the tables in the official reports of the meteorological office for the period.

TABLE 2.10: RAINFALL IN TIRUCHIRAPALLI IN DAYS AND IN INCHES, 1827-76

Month	Number of Days	Rainfall in inches
January	1	1.0
February	½	0.5
March	1	0.7
April	1	1.8
May	5	3.8
June	3	1.3
July	3	2.2
August	7	4.4
September	8	5.3
October	11	7.8
November	9	5.2
December	6	3.1
Total	55½	37.1

Source: Henry F. Blanford, *A Practical Guide to the Climates and Weathers of India, Ceylon and Burma, and the Storms of Indian Seas*, London: Macmillan, 1889, p. 329.

TABLE 2.11: RAINFALL IN MADURAI IN DAYS AND IN INCHES, 1827-76

Month	Number of Days	Rainfall in inches
January	1	0.7
February	1	0.4
March	1	0.6
April	2	2.0
May	5	2.8
June	3	1.6
July	3	1.7
August	7	4.7
September	8	4.5
October	14	8.7
November	9	5.1
December	5	2.2
Total	59	35.0

Source: Henry F. Blanford, *A Practical Guide to the Climates and Weathers of India, Ceylon and Burma, and the Storms of Indian Seas*, London: Macmillan, 1889, p. 330.

TABLE 2.12: RAINFALL IN COIMBATORE IN DAYS AND IN INCHES, 1827-76

Month	Number of Days	Rainfall in inches
January	1	0.3
February	1	0.1
March	3	0.6
April	4	1.8
May	10	2.6
June	9	1.8
July	9	1.3
August	8	1.2
September	6	1.2
October	15	5.7
November	13	3.4
December	6	1.1
Total	85	21.1

Source: Henry F. Blanford, *A Practical Guide to the Climates and Weathers of India, Ceylon and Burma, and the Storms of Indian Seas*, London: Macmillan, 1889, p. 330.

TABLE 2.13: THE FALL OF RAINS IN INCHES AT PALAYAMKOTTAI AND SHENKOTTAI REPORTED BY GENERAL CULLEN, 1844-8

Year	Palayamkottai	Shenkottai
1844	11½	24
1845	25½	43
1846	18	41½
1847	47	70
1848	15	32½

Source: *The Madras Journal of Literature and Science*, vol. XV, 1849, pp. 459, 463.

The English East India Company also had set up facilities to measure the rainfall in inches at Palayamkottai and Shenkottai in Tirunelveli region. These annual rainfall data was published in the *Madras Journal of Literature and Science.* Rainfall in 1823 at Chingleput was 26.62 inches and in 1832 it was 18.45 inches, as mentioned in the almanac of 1892 printed at the Asylum Press, Madras.

Further, to test the reality of the supposed periodicity shown, Strachey rearranged the series of 64 years' observations in a purely arbitrary manner, in cycles of 11 years, by drawing the actual observations at random one after another, and setting them down in succession till the whole exhausted. From three arbitrary sets of six cycles thus prepared, the mean cyclical difference averaged 10.9, 11.2 and 11.6. These results again indicated that by adopting the actual sequence of the observed quantities of rain instead of taking them at random, it produced no material effect on the mean cyclical differences or any such tendency to a diminution in their numerical value as necessarily accompanied a true periodical element. It was, moreover, important to bear in mind that the mere circumstance of any series of cyclical means showing a single maximum and single minimum gave no more real indication, that such a result was a truly periodical feature than would be supplied by the appearance of two or more maxima and minima. The law of periodicity, if it existed at all, could only be inferred by the facts indicated by observation; and it was obviously to argue in a circle, first to assume a cycle on which to work, which should give a single maximum and minimum, and then to infer that there was true periodicity because of the single maximum and minimum. The test of the periodicity was in truth to be sought altogether outside of the particular values of the successive cyclical means. It was, of course, manifested that a complication of periodical elements might so mask one another as to prevent positive results being obtained by such an examination of the cyclical means and differences as Strachey had made.[128]

The whole scope of Strachey's argument was negative, and it necessarily led to the conclusion that the cyclical variations shown in Hunter's study from the mean values were so great as to show that any apparent regularity or tendency to a maximum in one part of the 11-year cycle, and a minimum in another, had no real weight, and that there was no proof of greater tendency to periodicity in the 11 year means than in the original isolated observations. An objection might possibly be made to the effect that the sunspot period was not exactly a cycle of 11 years, and that a better result might be obtained by a comparison of the observations corresponding to the known periods of maximum and minimum sunspots, without reference to any special length of cycle. Strachey gave tables to show the results thus arrived at, the figures representing differences from the mean rainfall for the entire 64 years. These results were also negative. The maximum period could not be said to be indicated at all. The minimum showed a diminished mean cyclical deviation, but this was due to the years farthest from the supposed minimum epoch; and if the three central years alone were reckoned, the result was much the same as obtained from the 11-year cycles.[129] Thus, we find the flaw possibility that lay in the statistical model of rainfall that the individuals used to make predictions.

The *aarus* in Tamil country did not throw up lands for cultivation on large scale. The terms *aatru neer* and *murai neer* appear meaning the right to limited and full use of water, respectively, from a water source and also particularities such as that the transferee should enjoy the right of making an *anaikattu* to lead water to his field in the eighteenth century.[130] The English East India Company administration noted the decrepit condition of *vaykkals*, *kaals*, *kulams* and other irrigation works. At first, the Kaveri river system

TABLE 2.14: AVERAGE MONTHLY RAINFALL IN INCHES AT MADRAS, CUDDALORE, TIRUCHIRAPALLI, AND MADURAI FOR 1875-85

Month	Madras	Cuddalore	Tiruchirapalli	Madurai
January	1.0	1.0	1.0	0.7
February	0.3	0.3	0.5	0.4
March	0.4	0.4	0.7	0.6
April	0.6	0.9	1.8	2.0
May	2.2	1.5	3.8	2.8
June	2.1	1.4	1.3	1.6
July	3.8	2.3	2.2	1.7
August	4.4	5.1	4.4	4.7
September	4.7	4.7	5.3	4.5
October	10.8	8.1	7.8	8.7
November	13.7	14.1	5.2	5.1
December	5.1	5.7	3.1	2.2
Year	49.1	45.9	37.1	35.0

Source: Henry F. Blanford, *A Practical Guide to the Climates and Weathers of India, Ceylon and Burma, and the Storms of Indian Seas*, London: Macmillan, 1889, p. 73.

was surveyed by government officers and engineers, and efforts were made to desilt the older *vaykkals* and construct new *kaal* systems.[131] In 1851, concurrent to the *aaru* improvement works, it was felt that the colonial government should charge a fee for water, separate to that of land revenue.[132] In order to supervise and coordinate the works, the Public Works Department for the first time in British India was set up in 1852 in Madras. The colonial state had primarily aimed to improve agricultural productivity, strengthen the revenue extraction mechanism and directly contribute to increasing monetary amounts. It was at this point that discussions began with respect to invigorating a regime of colonial water rights and control. Initially, it was planned to legally compel landed proprietors to contribute monetarily to the construction and upkeep of public works in the Madras Presidency. The second method was to revive and revitalize the ancient practice of *Kudimaraamath* (the rendering of compulsory free labour) by peasants for repair of embankments, drainage works and other repairs pertaining water.[133] Thereafter, it was planned to levy a fee for usage of water resulting from improvement works. The state had thus wished to assert its proprietary rights over water, especially with reference to specific benefits from *aaru* improvement works, which it had undertaken.

THE BRITISH COLONIAL CONTROL OVER WATER: LEVY OF WATER CESS IN TAMIL COUNTRY IN 1865

The Government of Madras, therefore, enacted the Irrigation Cess Act in 1865 and it was designed to extract revenue from water as separate as from that of land. Under this Act, a *zamindar* (holder of zamin), *inamdars* (holders of inam) and *mirasdar* (holders of mirasi), would have the right to water, according to the settlement the colonial government had made with that particular landholder at various points from the eighteenth to the mid-nineteenth century. Secondly, though the act allowed for the levy of a water cess, it refused to impose a blanket rate for all the use of water made available through government-constructed improvement works in the Madras Presidency. Instead, the district collectors could issue rules for every particular *anaikattu*, under which the water rates would be fixed.

K.V. Naidu, a distinguished lawyer in Madras at the time, objected saying that under the provisions of the Irrigation Cess Act of 1865, the government was reduced, time and again, in court to being 'a mere riparian proprietor under the English law', as opposed to a central managerial authority.[134] However, the matter was dragged on for a long time and nothing substantial came out of it.

It is necessary to compare the medieval practice of levying water cess and collection in Tamil country. Under the Pallava rulers, one *kaadi* of paddy was collected as water cess for one *patti* of the land irrigated by the *yeri* in Porpandal area.[135] Seventy *kaadi* of paddy was collected for one patti of land in Perumanur area for the upkeep of the *kulam*.[136] Another record mentions 150 *kaadi* of paddy collected at Parudir for the upkeep of the *kulam*.[137] Thus, water cess was mainly collected in kind and it varied from area to area.

Medieval Chola records also mention payment of a water cess which varied from region to region. We find mention of two *naali* of paddy per *kalam* was assessed as the rate of water cess.[138] Another record refers to one kalam of paddy on every *maa* of land irrigated from the *kulam* was imposed at Madhuranthakam.[139] One *kurni* of paddy on every *maa* was collected for desilting the village *kulam* and its maintenance at Tiruchirapalli.[140] One *pathakku* of paddy was collected on every *maa* of land for desilting the local *kulam* in Pondicherry area.[141]

Attrukaal vetti muttai al, *Aatrukulai* and *Attrupaatam* refer to the three types of taxes collected on river-water use. *Senneer amanji, senneer poduirai, senneer-vennir* and *senneer vetti* have been referred to as the four types of taxes collected on drinking water. *Yeri aayam, Yeri meen kaasu, Yeri meen paatam, Yeri neer kovai, Yeri vaay padikkaval* and *Yeri patti* have been mentioned as the six types of taxes levied for the maintenance of the *yeris.* However, there are nine types of taxes on water for maintenance and repair of irrigation works in general, such as *Neer amanji, Neer aarambam, Neer aaru, Neerk kirai, Neerkiya vilai, Neer kuli, Neernilai Kaasu, Neer vilai kuli* and *Neer nilai.* It is known from records thus that 22 types of taxes on water were collected from landowners and tenants in the medieval period. Hence, the British also aimed to follow the practice but the mode of payment was in cash.

It may be important to note the *kulams* and reservoirs built by the medieval rulers in Tamil country for storing rain water had two distinct advantages. They are the *aaru* embankments and *karais* of the *vaykkals.* We find dressed stones being used in place of the traditional rubble and laterite material in the Pandya kingdom. The stones were laid with precision using the string-line-technique which is referred to in the inscription as *nulitteruvitta* (string line).[142] *Kartthumpu* (stone sluices) were installed at suitable places to regulate the supply of water from the *kulam.*

The English East India Company officials who visited many *kulams* and reservoirs in the modern age expressed their opinion and this is significant. Cox who visited the Mahendravadi kulam wrote that the date of its construction was not known but it irrigated lands that were at an approximate distance of seven or eight miles.

He mentioned that the *karai* of Mahendravadi kulam was enormously high and might be restored to its original height in which case a great extent of land could be brought under irrigation. Though water-spread was not extensive as that of the Kaveripakkam kulam, the depth of water was much greater and the supply lasted for 15 months. He also opined that the Mahendravadi kulam was originally larger than that of Kaveripakkam.[143]

CONCLUSION

Rainfall was the central regulator of agricultural activity in Tamil country and so the long-term policy of the people had mainly focused on the development of irrigation. They also tried to reorganize production by bringing more uncultivable lands under their own personal supervision. There were no efforts on the part of the state to bring the irrigational facilities of the land under their direct supervision. Neither was there any kind of mechanism or state apparatus to exercise control over the water system under the Cholas, the Pandyas and the Vijayanagara rulers. The prosperity of agriculture in the period depended to a large extent on the facilities provided for irrigation more especially by the local *sabhas* and individuals who realized the importance of securing adequate water supply. People, therefore, tried to get water by four different ways such as by lift irrigation, from bailing, from the channels cut from *aarus* and by constructing ponds and *kulams*, and *kinarus. Kaals* and *kulams* were the two main systems followed in irrigation. Steps were taken to ensure that the surplus water that flowed in the *aarus* were diverted and reached the newly excavated *yeris* and *kulams* at convenient places. In this way, we find a chain of *yeris* and *kulams* along the course of *aarus* contributing significantly towards the development of agriculture. The *sabhas* took special interest to promote irrigation facilities by constructing various water bodies of different capacity for storing water.

Rainfall is the most important element in the study of climate of the Tamil region. The summer monsoon did not give much rain to the east coast of India as the moist winds exhausted most of the moisture by giving heavy precipitation on the west coast. The rainfall in the dry season was scanty. Coastal Tamil Nadu received its heavy showers/rains in the north-east monsoon from October to November. Cyclonic rainfall also occurred during November and December. The medieval inscriptions, the so-called para-meteorological records, usually mention events, such as floods, storms, heavy rain and drought. These are always relative, as meteorological parameters were not standardized or measurement instruments (e.g. rain gauge and thermometer) did not yet exist in the medieval period and thus they were dependent on the observer. Documents did not usually report favourable or reasonable weather conditions. Instead, they emphasized extreme and extraordinary weather conditions and weather-related phenomena. The inscriptions usually described the main characteristics of the weather or phenomenon in just a few words. The lack of climatic data from one year or decade might not mean favourable climatic conditions. The climatic conditions might have been considered not to be important events and thus were never written down. A study of rainfall statistics for the medieval period between the ninth and seventeenth centuries is not possible as there are no archival and scientific records. We find in the inscriptions of the medieval period that rainfall had been abundant, sometimes enough, in some places scanty, besides less and no rains for some years. Thus, the distribution of rainfall varied enormously from region to region in Tamil country through the ages.

The diaries in Tamil, maintained by the natives in Pondicherry, in the eighteenth century mention monsoon rainfall that could be classified into some categories: deficient (less than 90 per cent of the long period average; LPA), below normal (90-6 per cent), near normal (96-104 per cent), above normal (104-110 per cent), and excess (above 110 per cent), as in present-day standards. In the absence of using instruments in the past, it is not possible to make fine calculations. Muthu Vijaya Thiruvengadam Pillai of Pondicherry provided information on rainy days and the nature and actual condition of rains for AD 1794-5. The rainfall shows that rains started in Pondicherry on 5 October 1794, but in the following year (AD 1795) it started on 31 October, indicating a delay of 25 days in commencement of the rainy season. Rains stopped in Pondicherry on 10 December 1794 but in 1795 it stopped on

30 December (after 19 days). Thus, the rainy season was extended by six days between these two years.

In Europe, Edmund Halley mentioned in AD 1686, that the primary cause of the monsoon was the differential heating between ocean and land and the monsoon was considered to be a gigantic land-sea breeze. He associated monsoon with a system special to the monsoonal region and the intensity of the monsoon was directly related to the land-ocean temperature contrast.[144] Europeans in Tamil country began to engage themselves in the study of monsoons and noticed the south-west monsoon as a life-giving phenomenon which released maximum rainfall on the landmass. Every year, between June and September, moisture-laden winds blew in from the Indian Ocean. The spatial and temporal variations of the monsoon had been so vast that it befuddled the missionaries and Company servants. They tried to fully understand how the clouds developed during the monsoon, how they were born, how they died out and how the monsoons began to change.

Monsoon rains in medieval Tamil country are mentioned to have been only during the months of mid-October and November. The south-west monsoon always lasted in the medieval period between June and September. It may be said that forests regulated and influenced the climate and availability of water since, scientifically speaking, half of the rainfall stemmed from the evapotranspiration from the vegetation and the rest rose from the evaporation of the sea.

Europeans on the Tamil coast understood that a year in the Tamil calendar was divided into four seasons, namely (1) a cool dry season with north-easterly winds from mid-December to mid-February; (2) a hot, dry season with south-easterly and later south-westerly winds from mid-February to May; (3) the south-west monsoon season from June to September; and (4) the north-east monsoon season from October to mid-December. They also learnt that the land and sea breeze systems tended to modify or disrupt the monsoon system. The early modern European documents of rainfall in Tamil country provide climatic data that differ quantitatively and qualitatively. We get the precise details of rainfall in Tamil country in inches or millimeters. Hence, comparison of rainfall is possible for years and periods.

NOTES

1. G.U. Pope, *The Sacred Kural of Thiruvalluva Nayanar*, Oxford: Oxford University Press, 1886, Tirukural no. 737.
2. U.V. Swaminatha Aiyar, ed. *Puranaanuru*, Madras: Swaminatha Iyer Library, 1971, song no. 18, song no. 117.
3. Archaeological Survey of India, *South Indian Inscriptions*, [henceforth *SII*] Madras & New Delhi: Archaeological Survey of India, 1890 to 1990, vol. XXIII, no. 580.
4. T.V. Mahalingam, *Administration and Social Life under Vijayanagara*, Madras: University of Madras, 1940, p. 52.
5. K.A. Nilakanta Sastri, *Studies in Cola History and Administration*, Madras: University of Madras, 1932, p. 36.
6. *Annual Report on Epigraphy* [hereafter *ARE*], (Including Indian and South Indian epigraphy from 1881 to 1945, for the years 1887-1981, Government of India, Madras, no. 110 of 1925.
7. V.R.R. Dikshitar, 'A History of Irrigation in South India', in *Journal of Indian Culture*, 1945, pp. 71-2; K.A. Nilakanta Sastri, *The Colas*, Madras: University of Madras, 1937, p. 262.
8. Venkatasamy Nattar, ed., *Silapathikaaram*, Tirunelveli: South India Saiva Siddhanta Pathippu Kazhagam, 1992, canto XI. See also Haripada Chakarabarti, 'History of Irrigation in Ancient India', in *Proceedings of the Indian History Congress*, 32nd session, Jabalpur, Delhi: Indian History Congress, 1971, pp. 150-7.
9. F.E. Morgan, *Irrigation in Southern India*, ed., Somerset Payne London: Foreign and Colonial Compiling Publication Company, 1914-15, p. 285.
10. *ARE*, 39 of 1897.
11. *Epigraphia Indica* [hereafter *EI*], Calcutta & Delhi, XXII vols., Calcutta & Delhi, Government of India, 1892-1978 , vol. XII, p. 343.
12. *ARE*, 426 of 1924.
13. *SII*, vol. VII, no. 51.
14. S. Jeyaseela Stephen, 'Most buried treasure: Letters of the Portuguese Jesuits from Tamil Country in South India during the seventeenth Century', in *Hispanic Horizon*, vol. XVIII, no. 20, 2001, pp. 76-84, see p. 79.
15. *ARE*, 304 of 1909.
16. Archivum Romanum Societatis Iesu (hereafter ARSI), Rome MSS, Goa 56, Letter of Fr Pierre Martin to Le Gobien in December 1703, in *Lettera Annua della Missione del Madure*, December 1703.
17. N. Smith, *A History of Dam*, London: Orient Longman, 1971, p. 189.

18. K. A. Nilakanta Sastri, *The Colas*, reprint, Madras: University of Madras, 1984, p. 303; Kallathan yeri is mentioned in line six in *ARE* 183 of 1895; *Marai perari yean nam peral vetti kulathukkum* is found in *ARE* 122 of 1894; Yeri katti is recorded in *ARE* 122 of 1894.
19. *SII*, vol. XIV, no. 26.
20. T.N. Subrahmanyam, *Pandiyar Seppedugal Pathuu*, Madras: Thamizh Varalaatru Kazhagam, 1967, pp. 95-115; *Ennirntaka tatakankal* is mentioned in line 123.
21. *SII*, vol. II, part III, no. 106.
22. K.R. Srinivasa Aiyar, *Inscriptions of Pudukottai State* [hereafter *IPS*], Pudukottai: Pudukottai State, 1929, no. 478.
23. *SII*, vol. VI, no. 348; *ARE*, 1904-5, pp. 131-45.
24. *SII*, vol. III, no. 99.
25. *ARE*, 322 of 1923; *Yerivaariya perumakkal*, the members of the tank committee, are mentioned in *ARE*, 65 of 1898.
26. *ARE*, 178 of 1902; 192 of 1909.
27. *ARE*, 156 of 1942-3; T.M. Srinivasan, *Irrigation and Water Supply: South India 200 BC-AD 1600*, Chennai: New Era Publications, 1968, pp. 30-2.
28. *SII*, vol. II, part III, no. 106.
29. Y. Subbarayulu, *Studies in Cola History*, Chennai, 2001, pp. 93-4; James Heitzman, *Gifts of Power: Lordship in an Early Indian State*, Delhi: Oxford University Press, 1998, p. 177. *Vetti mutal* and forced labour is mentioned in *SII*, vol. VII, no. 411.
30. *ARE*, 152 of 1934-5.
31. *ARE*, 215 of 1919.
32. *SII*, vol. XIV, nos. 43, 44.
33. Rajan Gurukkal, *The Agrarian System and Socio-Political Organization Under the Early Pandyas, 600-1000,* Unpublished Ph. D Dissertation, Jawaharlal Nehru University, New Delhi, 1984.
34. *ARE*, 285 of 1965-6.
35. *SII*, vol. XIV, no. 78.
36. *SII*, vol. XIV, no. 91.
37. *SII*, vol. XIV, no. 19.
38. *SII*, vol. XIV, no. 86.
39. *ARE*, 424 of 1909.
40. Purushothaman vaykaal mentioned in line 3, *ARE*, 183 of 1895; Sri Mahdhava vaykaal mentioned in line 3, *ARE* 183 of 1895; Parameswara vaykaal mentioned, *ARE* 186 of 1895; Kothukula vaykaal, Iruumudichola vaykaal, Ponngal vaykaal, Chandrasekara vaykaal are mentioned, *ARE* 173 of 1895.

41. *EI*, vol. XXXVIII, part I, pp. 27-32.
42. *ARE*, 33 of 1961-2.
43. *SII*, vol. XIII, no. 13; *SII*, vol. XIII, no. 163.
44. *ARE*, 200 of 1934-5.
45. *IPS*, no. 573.
46. *IPS*, no. 582.
47. *ARE*, 88 of 1931-2.
48. *ARE*, 152 of 1934-5.
49. *SII*, vol. XIV, no. 17.
50. *SII*, vol. XIV, no. 400.
51. N. Kasinathan and R. Nagaswamy, eds., *Kanyakumari Kalvettugal*, 5 vols., Madras: Tamilnadu State Department of Archaeology, 1968-79, nos. 212 & 224.
52. *ARE*, 71 of 1890.
53. *SII*, vol. I, p. 86.
54. *SII*, vol. VIII, part III, p. 437.
55. S. Jeyaseela Stephen, *The Coromandel Coast and its Hinterland: Economy, Society and Political System, 1500-1600*, New Delhi: Manohar Publishers, 1997, p. 148.
56. *ARE*, 282 of 1929-30.
57. *ARE*, 14 of 1909; *EI*, vol. XXVIII, Part I, pp. 27-32; Rajan Gurukkal, 'Aspects of the Reservoir System of Irrigation in the Early Pandya State', in *Studies in History*, vol. II, no. 2, 1986, pp. 155-62, see p. 156.
58. *ARE*, 54 of 1905; Kasinathan, *Kanyakumari Kalvettugal*, no. 224.
59. *ARE*, 295 of 1909.
60. *SII*, vol. XIV, no. 40.
61. Rajan Gurukkal, The Agrarian System, p. 160.
62. *ARE*, 79 of 1928-9.
63. *ARE*, 351 of 1965-6.
64. *ARE*, 27 of 1927.
65. Noboru Karashima, *South Indian History and Society: Studies from Inscriptions, AD 850-1800*, Delhi: Oxford University Press, 1984, p. 48.
66. Ibid.
67. *SII*, vol. XIV, no. 26.
68. *SII*, vol. XIV, no. 44.
69. *Nikkurngudi vaykaal paarai nikki* is mentioned. *ARE*, 183 of 1895.
70. *EI*, vol. XV, p. 72.
71. *SII*, vol. VIII, nos. 66, 67; T.N. Subrahmanyam, *Pandiyar Seppedugal Paththu*, pp. 95-115; 212-13 & 219.
72. *SII*, vol. II, no. 73.

73. *ARE*, 7 of 1926.
74. C. Jegannatha Charier, ed., *Erelupadu by Kambar*, Madras, 1967, verse 39, p. 52.
75. S. Jeyaseela Stephen, *Letters of the Portuguese Jesuits from Tamil Countryside, 1666-88*, Pondicherry: Institute for Indo-European Studies, 2001, p. 138.
76. *ARE*, 394 of 1912; Transcript lines 4 and 5 runs as follows: *kinarum vetti kalaniyum thiruthi payir seidu kollakkadavaragavum*. The word *kinaru*, *Kinarrtu nir pirati* is found very frequently in the Tiruvannamalai region; P.R. Srinivasan & Marie Louise Reniche, *Tiruvannamali: Inscriptions*, Pondicherry: Ecole Francaise d' Extreme-Orient, 1990, nos. 375, 379, 392, 395.
77. *ARE*, 103 of 1921.
78. *ARE*, 118 of 1902.
79. *SII*, vol. XIV, no. 56.
80. *ARE*, 394 of 1912.
81. *ARE*, 388 of 1912.
82. *ARE*, 387 of 1912.
83. Sadhu Subrahmanya Sastry, *Tirumalai-Tirupati Devasthanam Epigraphical Series* [hereafter *TTDES*], 6 vols., Madras: Tirumalai-Tirupati Devasthanam, 1931-8 vol. III, pp. 1-3.
84. *TTDES*, vol. IV, pp. 74 & 105-6.
85. *TTDES*, vol. III, pp. 103-8.
86. *TTDES*, vol. IV, pp. 150-1.
87. M. Rama Rao, *Telugu Inscriptions of Andhra Desa*, Tirupati: Sri Venkateswara University, 1969, vol. I, part I, no. 891 (1592); *ARE*, 188 of 1916 (B).
88. Burton Stein, *Peasant State and Society in Medieval South India*, Delhi: Oxford University Press, 1980, p. 426.
89. *ARE*, 179 of 1913 (1533); 94 of 1913; 388 of 1916 (B).
90. *ARE*, 145 of 1924.
91. *ARE*, 194 of 1921; 64 of 1942.
92. *ARE*, 183 of 1921-2.
93. T.N. Subramanian, *South Indian Temple Inscriptions*, Madras [hereafter *SITI*], III vols., Madras: Government Oriental Manuscripts Library, 1953-7 vol. I, p. 342 (1549), p. 339 (1541). The Kanchipuram inscriptional text runs '*Vittavan Vilukkadu Nalil Onrukku*'.
94. *TTDES*, vol. IV, pp. 264-5.
95. *TTDES*, no. 72, 74 and 93; vol. V, no. 88; vol. III, nos. 157, 159.
96. *ARE*, 54 of 1934-5.

97. *ARE,* 119 of 1931.
98. *ARE,* 251 of 1904.
99. *SITI,* vol. I, pp. 43-4.
100. *ARE,* 177 of 1941-2.
101. *TTDES,* vol. II, pp. 329-32.
102. N. Venkataramanaya, *Studies in the History of the Third Dynasty of Vijayanagara,* Madras: University of Madras, 1939, p. 187.
103. *TTDES,* vol. I, no. 224.
104. *ARE,* 16 of 1938-9.
105. *ARE,* 9 of 1922.
106. *ARE,* 265 of 1936-7.
107. S.M. Kamal, *Sethupathi Mannar Seppedugal,* Ramanathapuram: Sharmila Pathippagam, 1992, p. 112.
108. S.M. Kamal, *Sethupathi Mannar Varalaaru,* Ramanathapuram: Sharmila Pathippagam, 2003, pp. 73-4.
109. Thomas Bowrey, *A Geographical Account of the Countries Round the Bay of Bengal, 1669-79,* Cambridge: Cambridge University Press, 1905, p. 132.
110. British Library [hereafter BL], London, Oriental and India Office Collection [hereafter OIOC], MS Original Correspondence, no. 4215.
111. H. Dodwell, ed., *The Private Letter Books of Joseph Collet, Sometime Governor of Fort St George,* New York, John Murray, 1933, pp. 121, 184.
112. J.E. Geister was born in Berlin, studied at Jena and Halle and was ordained in Wernigerode in 1731. He arrived at Madras in 1732, went with Sartorius to Cuddalore in August 1737, and carried out missionary work there. When Fort St. David at Cuddalore was occupied by the French in 1743, he moved to Madras. He died on the voyage back to Europe in 1746. See Johan Ferdinand Fenger, *History of the Tranquebar Mission: Worked out from Original Papers,* trans., Emil Francke, Tranquebar: Evangelical Lutheran Mission Press, 1863, p. 314.
113. Archiv der Franckeschen Stiftungen, Halle [hereafter AFSt], AFSt/M 2 B 2:2, 'Wind and Weather Observations 1732-7, Madras, 20 October 1732'; see also *Wurzburger Geographische Arbeiten,* vol. 80, 1991, pp. 45-86; see Buchard Brentjes, 'Meteorlogische beobachtungen des 18.Jahrhunderts in Madras', in Burchard Brentjes and Hans-Joachim Peuke, *Wissenschaftsbeziehungen zwischen Halle und indien in tradition und Gegenwart,* Halle: Martin-Luther-Universitat Halle-Wittenberg, 1987/44 (142), pp. 27-38.
114. R. Walsh, R. Glaser and S. Militzer, 'The Climate of Madras during the Eighteenth Century', in *International Journal of Climatology,* vol. 19, issue 9, July 1999, pp. 1025-47, see p. 1028.

115. Ibid., p. 1031.
116. Ibid., p. 1032.
117. *Philosophical Transactions of the Royal Society of London*, vol. LXX, 1778, pp. 182-92.
118. *The Madras Journal of Literature and Science*, vol. 1, New Series (no. XVIII, Old Series), October 1856-March 1857, pp. 205-8, see p. 205.
119. John Goldingham, 'Results of Meteorological Inquiries Made at Madras', in *Transactions of the Royal Asiatic Society of Great Britain and Ireland*, vol. 3, no. 3, 1834, p. xxx.
120. Ibid., p. xxi.
121. Ibid.
122. *The Madras Journal of Literature and Science*, vol. 1, New Series, p. 205.
123. Ibid., p. 206.
124. Ibid., p. 208.
125. General Richard Strachey (1817-1908) returned England in 1871. He became the president of the Royal Geographical Society in London. See, Strachey 'On the Alleged Correspondence of the Rainfall at Madras with the Sun-Spot Period and on the True Criterion of Periodicity in a Series of Variable Quantities', in *Proceedings of the Royal Society of London*, vol. 26, 1877, pp. 249-61, see p. 249.
126. Ibid., p. 250.
127. Ibid., p. 251.
128. Ibid., p. 252.
129. Ibid., p. 253-4.
130. Noboru Karashima, *South Indian History and Society*, p. 173.
131. Baird Smith, *The Cauvery, Kistnah, and Godavery: Being a Report on the Works Constructed on these Rivers for the Irrigation of the Provinces of Tanjore, Guntoor, Mausilpatam, and Rajahmundry in the Presidency of Madras*, London: Smith, Elder & Co., 1856, p. 146.
132. BL, OIOC, V/23/192, *Collection of Papers Relating to Control of Government over Water in the Madras Presidency, 1851-59*, p. 3.
133. BL, OIOC, L/PJ/6/107.
134. Kurma Venkata Naidu, *The Revised Irrigation Bill: Being a Reprint of a Series of Articles Contributed to the Indian Patriot*, Madras: Loganadham Bros., n.d., p. 4.
135. *ARE*, 402 of 1923; 258 of 1968-9.
136. *ARE*, 38 of 1934-5.
137. *ARE*, 252 of 1921.
138. *ARE*, 369 of 1937-8.
139. *ARE*, 1912 of 1919; 38 of 1934-5.

140. *SII*, vol. XIII, no. 165.
141. *SII*, vol. VII, no. 805.
142. *SII*, vol. XIV, no. 44.
143. Arthur Frederick Cox, *A Manual of the North Arcot District in the Madras Presidency*, Madras: Government Press, 1895, vol. II, p. 305.
144. Edmund Halley, 'An Historical Account of the Trade Winds, and Monsoons Observations in the Seas between the Tropics with an Attempt to Assign the Physical Cause of the Said Winds', in *Philosophical Transactions,* vol. 16, 1686, pp. 153-68.

CHAPTER 3

The Famine and Drought in Tamil Society, Ninth-Nineteenth Centuries

Ecstasy and agony seem to come in tides to peasants because of the rain. Lack of or inadequate rainfall, leading to scarcity of water, has offered moments of agony. After years of dearth and drought, farmers were very happy when it rained well. Drought meant absence of drinking water or groundwater. Surface water was a viable and relevant source as a provision of drinking water. Hence, surface water storage in reservoirs in Tamil country has played a very significant role throughout centuries.

THE FAILURE OF MONSOONS IN THE MEDIEVAL PERIOD

When monsoons failed in Tamil country, it brought untold suffering to people owing to non-availability of water. It is mentioned that there was no rain continuously for 12 years during the rule of Raja Simha (AD 691-729) the Pallava king and thus a terrible famine broke out and continued in Kanchipuram. A Chinese source mentions how Vajara Bhoti, a Buddhist ascetic, went there and relieved the distress of the people by performing a miracle.[1] Dandin, the court poet of Raja Simha, narrates the consequences of this famine as follows: 'Family women were molested, the worship in the temples stopped, the granaries became empty, householders were driven out, honour was destroyed, rows of trees and garden were devastated, Kali was the sole monarch'.[2] We also learn that King Narasimha Varman II (AD 695-722) had appealed to ascetics to cause rainfall.[3] The commentator of Iraiyanar Ahapporul (eighth century) speaks of the absence of adequate rains for 12 years and the outbreak of famine in Pandya country which devastated both people and the state.[4] Thus, famines were generally caused due to lack of rain and continued drought prevailed in the early medieval period. The Pallava rulers prepared themselves for tackling famine by building a buffer-stock of the harvest through Panchavara Vaariyam.[5] Therefore, political and administrative measures had been taken but they often proved to be insufficient to tackle these challenges effectively.

The famine periods have been referred to as *kali yugam*, the period of bad times, in the inscriptions of Tamil country. The Velur Palayam copper plates of AD 852 (*Angirasa*, the sixth year of the Tamil calendar) mentions that the ruler governed the earth and he was trying to clear the dark ages.[6] The Bahur copper plate inscription in AD 877 (*Yevilambi*, the thirty-first year of the Tamil calendar) describes that the king ruled the earth according to *dharma* even in the *kali yugam.*[7] The Tiruvaalangadu plates of AD 1019 (*Siddharti*, the fifty-third year of the Tamil calendar) state that the people of the country considered their ruler Paranthaka Chola I to be Manu himself, reborn to establish his own laws which had degenerated in the *kali yugam.*[8]

The Chola records of the medieval period also mention frequent failure of the monsoons. There was a famine at Thirukarugavur in the Thanjavur region that broke out because of the failure of monsoon in AD 1019 and the crops had been very badly affected.[9] Owing to the failure of monsoon in AD 1121, the waters of Jagnnatha peraaru and Parantha peraaru had dried and as a result a severe drought had occurred.[10] In the months of Avani-Puratasai (August, September-October), there was failure of monsoon in AD 1160 at Panjai and Thirukadaiyur in the region of Thanjavur. As a result, crops were damaged and there was drought too.[11] The failure of monsoon again affected the people at Thirukkachur in Chingleput region in AD 1188. It states the failure of crops which finally led to severe famine too in the entire area.[12]

Monsoons failed in AD 1201 at Thiruppachuram in Thanjavur region and subsequently famine and drought plagued the region.[13]

The famine also continued in the following year (AD 1202) in Tiruvannamalai region.[14] It revisited Ayyampettai region in AD 1208 and it is mentioned that agriculture had suffered greatly.[15] The failure of monsoon is noted in AD 1215 in Thirukacchur region.[16] People of Thirumangalakkudi in Thanjavur area were affected due to the failure of rains in AD 1239 and a famine broke out.[17] The people in the village of Alangudi suffered grievously, owing to lack of adequate rains and resultant famine.[18]

TABLE 3.1: THE PERIODICITY AND FREQUENCY OF FAMINES IN THE LOCALITIES OF MEDIEVAL TAMIL COUNTRY, AD 1019-1396

Year AD	Name of the Place	Region	Year amongst the 60 Years of Tamil Calendar
1019	Tirukarugavur	Thanjavur	Siddharti, 53rd year
1054	Alangudi	Pudukottai	Jaya, 28th year
1121	Alangudi	Pudukottai	Parabhava, 40th year
1129	Tiruppur-Koviladi	Thanjavur	Saumiya, 43rd year
1131	Arakandanallur	South Arcot	Virodhikirudhu, 45th year
1153	Erumbur	Chidambaram	Srimukha, 7th year
1160	Panjai	Thanjavur	Vikrama, 14th year
1160	Thirukadaiyur	Thanjavur	Vikrama, 14th year
1188	Tirukachchur	Chingleput	Pilavanga, 42nd year
1201	Tiruppamburam	Thanjavur	Dunmathi, 55th year
1202	Tiruvannamalai	Tiruvannamalai	Dundhubi, 56th year
1208	Ayyampettai	Thanjavur	Vibhava, 2nd year
1215	Tirkkachchur	Chingleput	Yuva, 9th year
1226	Ilayangudi	Ramanathapuram	Viya, 20th year
1239	Thirumangalakudi	Thanjavur	Vikari, 33rd year
1241	Thirumangalakudi	Thanjavur	Pilava, 35th year
1390	Tiruppanangadu	North Arcot	Pramodhutha, 4th year
1391	Tirukkalar	Thanjavur	Prajothpaththi, 5th year
1396	Tamil Country	All over the area	Dhaathu, 10th year

Source: S. Jeyaseela Stephen, 'Rainfall and Irrigation in Medieval South India and the Socio-Economic Impact', in *Proceedings of the Indian History Congress*, 66th Session, Delhi, 2006, pp. 296-308; S. Jeyaseela Stephen, 'Water Resources and Management in the Rural Setting of Medieval South India', Peter Borschberg and Martin Krieger, eds., *Water and State in Europe and Asia*, Delhi: Manohar, 2008, pp. 105-20.

The Tamil calendar system has 60 years and so we take the years of famine as mentioned in the records. The Christian era (calendar having 100 years) is given for the purpose of convenience and easy reference. We are attempting at an exercise as to find out the frequency and the periodicity of the famines and droughts and how the failure of rainfall had a great impact. This would help us to understand the cycle of famines and droughts.

THE IMPACT OF FAMINES: PRICE ESCALATION AND PEOPLE ENTERING INTO SLAVERY

One of the symptoms of the failure of monsoon in Tamil country was the rise in the prices of food-grains. Medieval records reveal that crops were affected, resulting in the scarcity of rice and food. In Tiruppamburam area, the famine of AD 1201 was of a serious character and the scarcity of food-grains led to a high rise in prices. It is stated that paddy was sold at three *naali* for one *kasu*. This would mean that one-fifth measures of rice per *kasu* was prevalent.[19] Similarly in Tiruvannamalai area, rice was sold at one-fourth of a measure per *kasu*, owing to the famine in AD 1202 and there was a severe shortage of foodgrains.[20] These famines in various localities broke out on a different scale but prices of food-grains normally rose very high.

During the famine of AD 1201, many people had voluntarily become slaves. A *vellalan* (agriculturist) in the village of Tiruppamburam sold himself and his two daughters as slaves to the local temple for a sum of AD 110 *kasus* in order to escape death by starvation.[21] When the famine revisited in AD 1210, another poverty-stricken vellalan sold himself along with his wife to a *matha* (Hindu monastery/religious house). Those purchased as slaves were employed in the beginning but they were later resold.[22] Severe famines caused many hungry people to be sold as slaves, thus individuals sold their personal liberty for want of food. It is known that people had sold themselves continuously during famines in the thirteenth century in order to get food.[23] In AD 1390-1, when the famine broke out, paddy could not be purchased even at 10 *naali* per panam.[24]

THE ROLE OF MEDIEVAL TAMIL KINGDOMS AND FAMINES

A record states that during the famine in AD 1125 in Tiruvadi, the *sabha* had to sell some of the common land on account of the difficulty experienced in the payment of land tax to the king.[25] Yet another record refers to failure of crops and famine in Thirukachur, besides payment of land taxes in AD 1188 and 1215. The residents found it very difficult to pay taxes and thus were forced to take a loan from a nobleman in the neighbourhood.[26]

There are records that some concessions in the payment of taxes had been declared by rulers or even remission of some taxes during famines. Vira Pandya, the ruler in the eleventh century, permitted 50 per cent of *iraiyili* tax to be paid by the *nattar* (head of the local assembly) of Karraipattu paganeri due to the severe famine. In AD 1285, the residents of Sannavaanam were rehabilitated during the famine and so remission of taxes had been granted. *Sabha viniyogam* (the distributed levies collected by the *sabha*) from Kodanur Kulaskehara Chaturvedimangalam on the lands had been remitted during famine by the order of the ruler. Taxes such as *vettipattam, panjupili, sandivigh-rahapperu* and *pon vari* were also remitted during the famine. However, the land tax of *kadamai* and another tax *antarayam* had to be paid.[27] The medieval rulers had insisted on the payment of land taxes even during times of natural calamities since they needed revenue for the state exchequer.

No evidences attest to the existence of any famine policy which guided the governments. The medieval Cholas collected a levy called *panchavara.* It was a tax on one-fifth of the produce. It is unclear from records whether it was collected to provide against famines.[28] The role of the medieval state in relieving the famine distress in times of natural calamities and its reasons for not coming to the rescue of the people affected are shrouded in mystery.

ROLE PLAYED BY PRIVATE INDIVIDUALS AND VILLAGE ASSEMBLIES DURING FAMINES

In the absence of state machinery to fight against famines, the *sabhas* took active role. A record mentions that during the famine in AD 1054 in Alangudi when the village was affected, the *sabha*, expecting no help from the King, took the initiative of providing relief to the people. Since the villagers had no funds to purchase paddy of their own for consumption, the *sabha* secured a loan of 1,011 *kalanjus* of gold and 464 *palams* of silver consisting of temple jewels and vessels for the sustenance of the villagers. A portion of the money was also allotted for resuming cultivation. In lieu of this, the members of the *sabha* mortgaged 8¾ *velis* of village common land in favour of God.[29]

Another record mentions that during a severe famine in AD 1121, funds were raised for making available artificial methods of irrigation. On that occasion, the *sabha* of Jaganatha Chatuverdimangalam gave certain jewels to the temple of Pasuaptivaram Udiayar of the village in exchange for some common land.[30] Thus, we find that when a reduced amount of money was in circulation during the famine period, *sabhas* mortgaged the common lands and raised funds to alleviate the distress of the villagers.

Under the above-mentioned circumstances, the villagers and the *sabhas* as well as well-heeled private individuals seized the initiative to take water resources into their own hands or they were given the mandate by the local powers of the state to seize the initiative to store water for the dry season. It is reported that owing to the failure of monsoon, rivers had dried up in AD 1121. Hence, the people of the locality raised funds for irrigation.[31] The munificent activities of private individuals spending from their own pocket are also known. When there was a famine at Tiruvannamalai, two persons started relief works in the form of constructing an embankment to the Palar river. A fresh *kulam* was built with a *mathugul kumili* in the village in AD 1202, paying the labourers in gold and paddy as they desired.[32] It is noticed that famines did not leave institutions like the temples. There was no money for use. An individual called Tirunat Perumal, therefore, spent 400 *panams* for the temple during the famine in AD 1391.[33] As there was no organized relief measure during droughts by the rulers, the people themselves were forced to express their readiness to meet and solve the water problem through establishing man-made water bodies and they had planned well to manage it effectively.

RAISING CASH LOANS FROM TEMPLE TREASURIES DURING FAMINES

In addition to mortgaging the common lands, it is gleaned from a record that during the famine in AD 1153 in Erumbur when food-grains were scarce, a loan was arranged from the temple treasury by the *sabha* to tide over the distress. The rate of interest on this loan (60 *kasus*) was fixed at 2 *tuni* and *kuruni* of paddy per *kasu.*[34]

During the famine in Tirukkacchur in AD 1215, when crops had failed and troubles arose in connection with the payment of taxes, the members of the *sabha* took a loan of 15 *kasus* from an individual. As for the payment of interest, they gave him a piece of common land belonging to the village also in addition to agreeing to pay the taxes to the government on that land.[35]

We find that the *sabhas* of Alangudi and Erumbur villages had taken a loan from the local temple treasury in AD 1054 and 1153, respectively, during famines.[36] In Pallichandal in South Arcot area, we find that Neminatha, who was in-charge of the temple of Ganadratittaperumpalli, had declared some portions of Jambai village as an asylum for the distressed to settle down by an order.[37] The sthanattar had sold the jewels of the temple at Ilayangudi and utilized the sale proceeds to tide over a famine and 110 *panams* offered to the temple the following year was utilized for restoring the jewels of the deity.[38] Thus, temple officials also took active part in relieving the distress of the people during famines.

The residents of Kalahasti area in Tamil country faced a severe crisis with the failure of monsoons. The literary text, *Sri Kalahasti Mahatmyam*, written by Dhurjati, describes the pathetic famine conditions that resulted from the failed monsoon.[39] It is mentioned that there was no sign of thunder clouds. People had no choice but to change their food habits owing to insufficient crops. They were forced to eat an inferior quality of rice, large bulbous roots and tubers, and the pulp of date and palm trees. Some people tried to sow millets as it would give the yield within 60 days by watering the fields through the device of *ettam.* But even these water devices got spoiled and did not function as they had not been used for a long time. Men left the villages in search of food, leaving behind their wives and children, and many people died because of the severe food crisis. Some abandoned their houses and migrated to other areas. The situation became so deplorable that people resorted to cannibalism.[40] The neglect of traditional waterworks by the Vijayanagara state after the battle of Talikota in AD 1565 and by the subsequent kingdoms in Tamil country brought drought and famine. Small irrigation systems like *kulams* and ponds, masonry *anaikattus*, reservoirs, *yeris*, and maintenance of *kaals* had been neglected.

FAMINES ON THE COROMANDEL COAST, AD 1520-40

Portuguese sources mention the occurrence of famines on the Coromandel Coast. Tamil inscriptions also mention that famines broke out. It is mentioned that there had been no rainfall during AD 1520-40 and so cultivation of food crops failed. The price of essential commodities like rice and oil rose high. Portuguese traders in Mylapore helped the famine victims and so King Achyutadevaraya sent a letter expressing his thanks to the Portuguese captain of Santhome for rendering valuable help to the people.[41]

TABLE 3.2: FAMINE YEARS IN COROMANDEL, AD 1520-40

Year AD	Year amongst the 60 Years of Tamil Calendar
1520	Vikrama, 14th year
1531	Khara, 25th year
1532	Nandana, 26th year
1535	Manmatha, 29th year
1536	Durmukha, 30th year
1537	Yevilambi, 31st year
1538	Vilamba, 32nd year
1539	Vikari, 33rd year
1540	Sarvari, 34th year

Source: S. Jeyaseela Stephen, *Portuguese in the Tamil Coast: Historical Explorations in Commerce and Culture, 1507-1749*, Pondicherry: Navajothi Publishers, 1998, pp. 156-7.

FAMINE IN NAGAPATTINAM IN AD 1620 AND THE PORTUGUESE TRADE IN SLAVES

The Portuguese had settled at Nagapattinam and they conducted overseas trade. Private Portuguese traders were able to acquire supplies of slaves from the area because some natives had sold themselves into slavery due to famine. We have the testimony of a Hindu male named Mani (Meney in the record) of the Paraiyar caste, who admitted that in AD 1620 he had been forced to sell himself because of hunger and necessity, to Francisco Macedo, a Portuguese merchant and captain, and that a Portuguese broker named Joto Carin was behind this 'sale'. As a result, the vicar-general of the church of Our Lady of Immaculate Conception in Nagapattinam wrote that he had examined the slave named Mani of 25 years old and was fully convinced of the latter being a good captive; thereupon, the missionary had cleared the buyer's conscience and gave him a license to sell Mani as a slave exporting to Manila.[42] It is a fact that traders began to use such a process that was ostensibly meant to provide slaves with some legal recourse to their advantage.

FAMINES IN NAGAPATTINAM IN AD 1659, 1661, 1690-1 AND THE DUTCH TRADE IN SLAVES

The Dutch captured and occupied Nagapattinam in AD 1658. Famines had continued to bring misery to people in the region of Nagapattinam. Many came forward and sold themselves as slaves, because of which the Dutch slave trade flourished between 1659 and 1661.[43] An eyewitness account of Nagapattinam in 1661 vividly portrayed the situation. It said that the famine had worsened to such an extent that entire villages, hamlets and settlements were depopulated, leaving hardly anyone behind. Dead bodies were lying scattered in dry *kulams*, along common roads, indeed, even on the streets of Nagapattinam. Those who had survived until then looked more like lifeless corpses than living people. They cried along the fields and roads for a single meal to fill their empty stomachs before dying. In short, the account ended saying that the miseries and sufferings were such that no pen could describe the extremity of the situation.[44] Antão Proença the Jesuit also referred to the widespread famine in the region, which he attributed to the vagaries of war and the failure of the monsoon. He stated that mortality was so great that dead bodies were heaped up in great piles.[45]

There was failure of monsoon in 1660 and 1672 and so famines broke out in Thiruchirapalli and Thanjavur region.[46] During 1690-1, slave purchase came to a halt because the famine situation had improved. Later 5,000 slaves were purchased in 1694 when slaves were available for sale during famine at the rate of 1 *real* (10 *panams* equal to 1 real) per head at Nagapattinam.[47]

FAMINES IN PULICAT IN AD 1623, 1630, 1635, 1644, 1646, 1659, 1661, 1690 AND THE DUTCH TRADE IN SLAVES

Famine broke out on the Coromandel Coast in AD 1623. The Dutch who had settled at the port of Pulicat reported that rice was not available for export in June 1624.[48] The famine of 1630 that broke out on the Tamil coast helped the Dutch Company in Pulicat to purchase 100 slaves at a cheap rate. Again in 1635, when famine revisited the region, the Dutch purchased 91 slaves.[49] There were no slaves available after the famine in spite of the fact that the price of a slave had escalated between 18 and 20 reals.[50]

The Dutch in Pulicat could purchase slaves when famine revisited. Four ships that sailed on 3 March 1644 and another four ships that left on 5 October 1644 carried a total of 7,100 slaves.[51] The Dutch were able to acquire slaves because of the famine that continued to prevail till as late as 1646. A severe famine again had devastated Tamil country in 1647, chiefly owing to the failure of monsoon in the previous year. The famine also continued in the next year. The Dutch in Pulicat exported only 19 slaves on 21 May 1652 since the bad famine conditions were over. As a result, there were no slaves available for export.[52]

The famines of AD 1769-70 and 1781-3 resulted in extreme shortage of food-grains. The famine of 1781-3 followed a drought and it led to the establishment of charitable activities at Maniyak-karar Choultry in Madras where the poor were supplied free *kanji*

TABLE 3.3: FAMINE YEARS IN MADRAS, AD 1646-1779

Year AD	Year amongst the 60 Years of Tamil Calendar
1646	Viya, 20th year
1647	Sarvajitu, 21st year
1659	Vikari, 33rd year
1687	Prabhava, 1st year
1719	Vikari, 33rd year
1723	Sobhakirudhu, 37th year
1729	Saumiya, 43rd year
1730	Sadharana, 44th year
1734	Ananda, 48th year
1735	Rakshasa 49th year
1737	Pingala, 51st year
1747	Prabhava, 1st year
1781	Pilava, 35th year
1782	Subhakiritu, 36th year
1783	Sobhakirudhu, 37th year
1790	Sadharana, 44th year
1791	Virodhakrit, 45th year
1792	Paridhabi, 46th year
1798	Kaalayukthi, 52nd year
1799	Siddharthi, 53rd year

Source: Records of Fort. St. George, *Diary & Consultation Book, 1672 to 1756*, Madras: Government Press, 1910-50; H.D. Love, *Vestiges of Old Madras, 1640-1800*, III vols., London, 1913, reprint, Delhi: Asian Educational Services,1996; B.V. Narayanasamy Naidu, 'Famines in the City of Madras', in *Madras Tercentenary volume*, London: Madras Tercentenary Committee, 1939, pp. 73-87.

(rice gruel) every day. Consequent to the 1790-2 famine in Madras, the English banned rice export from Thanjavur to other places.

FAMINES IN PONDICHERRY IN AD 1687, 1708, 1747, 1760, 1788 AND THE ROLE OF THE FRENCH EAST INDIA COMPANY

The records of the French East India Company and the Tamil diaries maintained by the natives in Pondicherry during the seventeenth and eighteenth centuries provide substantial information and data on the famines and droughts that took place there. A famine broke out in AD 1687 in Pondicherry, causing scarcity of cereals and rice. François Martin, the agent of the French East India Company, arranged for the import of rice and grain by boats and these were then made available for sale.[53] It had been also reported that there was a drought in the French settlement owing to lack of rainfall in AD 1708 due to which famine also broke out in that year. It was so severe that the price of rice and other essential commodities rose exponentially. However, private charities from various sources came in a large measure to save and help the people of the town.[54] In order to ensure supply of food-grains to Pondicherry, the French Company suspended custom duties and levies for 36 days.[55] Famine appeared again in 1737 and there was an acute shortage of rice. The situation improved gradually when showers came in the end of January 1738. Nevertheless, the famine continued till October 1738 when one *garce* of rice was sold at a price between 80 and 100 *pagodas*.[56] Again, in 1759, there was no rain in the month of *Karthigai* (November-December) and so a disastrous famine broke out in Pondicherry that continued till 1760. Records explicitly state that the famines prevailed as far as Sethu (Ramanathapuram) in the south of the Tamil coast.[57]

The misery caused by the famine led to an increase in the prices of the commodities and goods by two-thirds. People were obliged to sell their possessions in order to feed themselves.[58] Most of the natives fled to safer places.[59] The French Company estimated that about 1,800 *piastres* were needed to feed the people of the garrison stationed in the settlement. The treasury had only 1,086 *pagodas*, 16 *panams* and 4 *kasus* at that time and this amount was just enough to pay the former debts of the French Company. Added to it was the heavy debt that the Company had incurred during the earlier famines from the various suppliers who had provided them with goods and were now demanding money.

Rice was sold at a high price of 50 *pagodas* per garce.[60] The Superior Council of Pondicherry decided to send ships to bring rice from Chandernagore and Balasore. Later, during 1747, when there was another famine, the governor fixed the selling price of several commodities. Paddy was ordered to be sold at three small measures or one large measure for one *panam*.[61] Betel leaf sellers were asked to sell a bundle of nine leaves per kasu.[62] Cotton was ordered to be sold at 26 *pagodas* per *candy*.[63] On 28 August 1748, Nalla-chetti, the *gumastha* of Hazarat-ullah Sahib, was permitted to receive grain from all the nearby villages and sell goods at a price he pleased to compensate the cost of his carriage. Further, all those who bought paddy in his name were also allowed to enjoy the same benefit.[64]

A severe famine broke out in Pondicherry in 1760 and so, once again, the price of commodities rose abnormally high. No dog or cat was left in the town and even rats were sold at Rs. 2 each. Soldiers were obliged to sell their possessions in order to feed themselves.[65]

According to later records, famine struck Pondicherry yet again in 1788 and essential commodities and foodstuffs again became unavailable since agricultural production was seriously affected.[66] In order to remedy the situation, the French Company ordered that no duty would be collected for one month on the foodstuffs brought to the Pondicherry market for sale.[67] This order, along with exemption of the customs duty, remained in force till May 1792.[68]

RELIGIOUS CONVERSIONS UNDERTAKEN BY FRENCH MISSIONARIES IN PONDICHERRY DURING FAMINES, AD 1723-1898

It has been gleaned from French missionary records that famines were major causes leading to large-scale conversion of many Hindus, including pariahs, to Christianity. The converts were the economically poor and were at the most disadvantaged end of the Hindu social hierarchy, the most vulnerable during famines. The missionaries were quite zealous in coming to the aid of those afflicted. Claude Barbier, the French Jesuit, wrote in AD 1723 that more than 200 adults were baptized. Such unending famines had provided the priests a new occasion to exercise their zeal. Several Hindus have found along with the preservation of their physical life, an assurance of the eternal life of the soul, through the baptisms that they received.[69] Again when there was famine during 1730s and 1740s, the number of converts to Christianity had increased.[70]

Poissevin, a missionary, wrote that in May 1747, when famine stalked Pondicherry, it created an occasion for the conversion of several Hindus.[71] The conversions during the 1740s were made possible to some extent by the liberal donations of Dupleix, the French governor of Pondicherry, and his wife, which were no doubt utilized to relieve famine-stricken people.[72]

According to missionary records, famines had led to conversions even during the nineteenth century. During the terrible famine of 1831-2 and that of 1876-8, French missionaries donated rice and other provisions along with other material help and so, many conversions also took place.[73] Before the famine of 1876-8, there were 14,200 Catholics in Pondicherry. But by 1886, there were 20,300. Marius J. Baulez, a missionary, said that the sudden increase in numbers was due to the great number of baptisms administered during the famine crisis.[74] The charity of the missionaries given during famines to the people led to the increasing number of conversions between 1876 and 1878.[75]

The new converts were helped by the missionaries through charity. Fr. Arul had baptized about 1,500 famine-stricken Hindus between 1866 and 1870 by distributing some rice and money.[76] Rains failed in 1875 and the following year was even worse. Poor people starved, and missionaries distributed foodstuff and spoke about the happiness of heaven and the action of providence. Thus, a spiritual gift was grafted with the material help they received from the missionaries. The people admired the Christian charity and this contact led them to accept the faith. François Jean-Marie Laouenan, Bishop of Pondicherry, wrote about these conversions and said that they gave 18 *kasus* per day to every baptized adult for their food. It was given only during the period of instruction. Some neophytes cost less, others more. In short, including the food and the salary of the catechist in-charge of teaching the prayers, the baptism of each Hindu cost about 10 francs on an average.[77] In 1898 alone, famine resulted in the conversion of 2,225 pariahs to Catholicism in the villages situated to the north of Pondicherry town.[78]

The missionaries received liberal donations from France to carry on with their work during the periods of famine. Indeed, sometimes such activities were held up due to the shortage of money flowing

TABLE 3.4: FAMINE YEARS IN THANJAVUR, AD 1743-1830

Year AD	Year amongst the 60 Years of Tamil Calendar
1743	Rudrodgari, 57th year
1765	Parthiva, 19th year
1776	Durmukhi, 30th year
1779	Vikari, 33rdyear
1781	Pilava, 35th year
1785	Visuvaavsu, 39th year
1797	Pingala, 51st year
1804	Rakthashi, 58th year
1811	Prajothpathi, 5th year
1827	Sarvajithu, 21st year
1829	Virodhi, 23rd year
1830	Vikrudhi, 24th year

Source: *Der Königl. Dänischen Missionarien aus Ost-Indien eingesandter ausführlichen Berichten, Von dem Werck ihrs Ams unter den Heyden, angerichteten Schulen und Gemeinen, ereigneten Hindernissen und schweren Umstanden; Beschaffenheit des Malabarischen Heydenthums, gepflogenen brieflicher Correspondentz und mundlchen Unterredungen mit selbigen heyden*, Teil 1-9, (Continuationen 1-108) Waiserihaus, Halle, 1710-72, vol. 5, no. 58, p. 1504; Georg Christian Knapp et al., *Neuere Geschichte der evangelischen Missions-Anstalten zu Bekehrung der Heiden in Ostindien aus den eigenhändigen Aufsätzen und Briefen der Missionarien erausgegeben*, Teil 1-8, (Stück 1-95) Waisenhaus, Halle, 1770-8/1795-1848; K.M. Venkataramayya, *Administration and Social Life under the Maratha Rulers of Thanjavur*, Tanjore: Tamil University, 1984, pp. 377-8; F.R.R.B. Hemingway, *Madras District Gazetteers: Tanjore*, Madras: Government Press, 1906.

from Europe.[79] Jean Fourcade, a missionary, remarked that famine was the most terrible and most heeded of the predications. People of low castes converted to Christianity as we had helped. There were no high castes who got converted.[80] These details illustrate the role of the missionaries during famine who brought about conversions in a peaceful and persuasive manner by offering necessary materials and psychological assistance voluntarily to needy and helpless Hindus, especially the untouchables. One can also really appreciate the help so rendered by them. Conversions also diminished considerably after the famine. The missionaries were, thus, undoubtedly a strong source of physical and psychological comfort to the needy during famines.

TABLE 3.5: FAMINE YEARS IN TIRUCHIRAPALLI, AD 1805-65

Year AD	Year amongst the 60 Years of Tamil Calendar
1805	Kurodhana, 59th year
1806	Akshaya, 60th year
1807	Pirabhava, 1st year
1837	Yevilambi, 31st year
1847	Pilavanga, 41st year
1865	Kurodhana, 59th year

Source: F.R.R.B. Hemingway, *Madras District Gazetteers: Trichnopoly*, Madras: Government Press, 1907; C.D. Maclean, *Manual of Administration of Madras Presidency*, Madras: Government Press, 1885, vol. II, pp. 70-1.

TABLE 3.6: FAMINE YEARS IN MADURAI REGION, AD 1804-99

Year AD	Year amongst the 60 Years of Tamil Calendar
1804	Rakthashi, 58th year
1805	Kurodhana, 59th year
1806	Akshaya, 60th year
1812	Angirasa, 6th year
1813	Srimukha, 7th year
1814	Pava, 8th year
1832	Nandana, 26th year
1836	Durmukhi, 30th year
1837	Yevilambi, 31st year
1847	Pilavanga, 41st year
1857	Pingala, 51st year
1865	Kurodhana, 59th year
1866	Akshaya, 60th year
1876	Dhaathu, 10th year
1878	Vehuthaanya, 12th year
1892	Nandana, 26th year
1899	Vikaari, 33rd year

Source: Tamil Nadu State Archives, Chennai, *Guide to the Records of the Madurai District, 1790-1835*; W. Francis, *Madras District Gazetteers: Madurai*, Madras: Government Press, 1906, pp. 96, 99, 114, 128, 132.

During the famine in AD 1743 in Thanjavur, there was the resultant scarcity of rice and prices reached a new peak. Circumstances were further aggravated as Tiruchirapalli was held under siege by a large army sent from Arcot, and its presence continued for several months. The troops needed provisions from Thanjavur. The high price of goods and famine worsened than it had been for years. It is mentioned that there was drought and it spread up to Ramanathapuram region in 1792.

Famines continued to occur due to lack of rainfall in Madurai region from 1812-57. On comparing the famine years at Madras and Madurai, we find only some years were common to both regions. In most cases, the famine years do not tally with each other. This proves that the rainfall pattern in Madurai, which is in the hinterland, was very different from Madras which is located on the sea coast.

TABLE 3.7: FAMINE YEARS IN MADRAS, AD 1812-97

Year AD	Year amongst the 60 Years of Tamil Calendar
1812	Angirasa, 6th year
1813	Srimukha, 7th year
1823	Subhanu, 17th year
1824	Tharana, 18th year
1825	Parthibha, 19th year
1832	Nandana, 26th year
1833	Vijaya, 27th year
1853	Piramadheesa, 47th year
1855	Rakshasa, 49th year
1866	Akshaya, 60th year
1876	Dhaathu, 10th year
1878	Vehuthaanya, 12th year
1889	Virodhi, 23rd year
1897	Yevilambi 31st year

Source: B.V. Narayanasamy Naidu, 'Famines in the City of Madras', in *Madras Tercentenary volume*, London: Madras Tercentenary Committee, 1939, pp. 73-87; Ravi Ahuja, 'State Formation and Famine Policy in Early Colonial South India', Sanjay Subrahmanyam, ed., *Land, Politics and Trade in South Asia*, Delhi: Oxford University Press, 2004, pp. 147-85.

SOLVING THE CRISIS OF FOOD-GRAIN SCARCITY DURING FAMINES BY THE ENGLISH COMPANY ADMINISTRATION IN MADRAS, AD 1782-1867

Severe famine and drought prevailed in Madras in AD 1782. Drought appeared again in 1823. Further, Madras and the northern areas were affected by severe drought in 1832. The English East India Company in the state had taken steps to prevent the increase of foodgrain prices besides scarcity during famines. The intervention of the Company state during famines in Madras is revealing. At first, the Madras government intervened in the grain trade only during the famine that broke out in 1782 and in 1791. When the people of Dindigul witnessed considerable distress during the famine in 1799, the collector of Madurai was permitted by the English Government of Madras to purchase grain on government account for the purpose of retail sale. It was also mentioned that if the famine were to grow more severe, he was to dispose of the grain he had in store at a loss in order to lower the general price of grain.[81] Five years later (in 1804), there was severe distress in Thanjavur and South Arcot areas. Prior to government sanction, the collectors of the two districts took it upon themselves to sell grain at low prices to people on government account, to prohibit grain export and import. Thus, it is gleaned that the state had intervened and shaped some sort of official famine policy.

The famine and drought had been interpreted by the Hindus as omens of God's will or as punishments of sins. People believed that famine and drought were outcomes of God's wrath and not natural phenomena. It is interesting to find that the collector of South Arcot district in consonance with Brahmin priests in Cuddalore had performed an *abishekham* ceremony to propitiate Varuna (the God of winds) and Indra (the rain God).[82] Science had been integrated with Hindu religion in Tamil country and Hinduism itself had been a nature worshipping religion. Professed in Bible (Book of Genesis), the Christian concept of the Europeans that man is superior to nature was different from the Hindus. The British government's involvement in such religious rituals did not occur in the later period. This indicates the growing reliance on the notion of 'scientific' which was increasingly juxtaposed by Europeans against a superstitious and unscientific native Hindu mind.

In 1804, the Government of Madras expressed objections to any interference with the grain market, unless absolutely necessary, but left the question of prohibiting exports to district collectors. As the famine dragged on in 1805, the Madras Board of Revenue laid down a set of principles upon which government policy should be conducted. Briefly, these were to prohibit any interference with the food-grains market either in terms of fixing prices, importing grain or selling it on government account, and to prohibit any embargo on exports. The Board of Revenue advocated the employment of people on public works, with payment in cash, as the only suitable means of intervention.

The East India Company State Government, while concurring with the proposals submitted by the Board of Revenue, authorized grain payments instead of cash due to the existence of government reserve. In 1807, the Government of Madras ordered that a fixed price was to be paid to merchants on the import of grain. Lord William Bentinck had, in fact, recommended that a bounty (or bonus) be paid instead of a fixed price as it was less dangerous to the government, but later conceded to his Council on the issue of a fixed price as ensuring a greater security of supply.[83] The following year, a committee was appointed to enquire into grain riots in Madras city, and it recommended that grain merchants be protected in selling by measure. They also authorized the employment of the able-bodied poor in the neighbourhood of Madras. The government also made considerable outlays on public works and considered establishing grain depots in the districts of Chingleput and North Arcot.[84] In December 1824, the Government of Madras realized that its estimation of famine distress had been woefully wrong, and it suspended duties on grain imports, and instructed the district collectors of Salem, North Arcot, South Arcot and Madras to open public works.

The Government of Madras also opened feeding houses for the relief of starving migrants in big towns during the famine in 1824. Thomas Munro made written statements affirming faith in non-interventionism as a famine policy. In response to a query on whether merchants were to be allowed to sell at their own rates, or whether the government would intervene and fix prices, the collector of the North Arcot district was told that he was to be guided by the true principles of political economy.[85] Munro wrote that the interference of the government on such occasions had very often been prejudicial, and he knew of no way in which it could be safe or useful, unless in suspending all duties on grains, thereby, giving perfect freedom to its transit by sea and land and securing the grain dealers from the violence of the people.[86]

The migration of the famine-stricken poor on a large scale to Madras city raised public awareness among the wealthy and rich. On 21 July 1866, a public meeting was held in Madras where a sum of Rs. 20,000 was collected for the relief of the starving poor. This meeting marked the beginning of systematic and joint efforts by local officials and private charities towards providing relief through a centralized system. The Government of Madras at that time resolved to contribute an equal sum to the relief fund and in future, similar sums to those raised in Madras would be thus contributed from the public treasury. These grants were spent towards the relief of the old, the infirm and the young.[87]

Simultaneously, employment for the able-bodied was planned to be provided on useful public works, although it has been unclear from records towards what type of works these were to be. The people were selected so as to allow them to be employed in gangs of moderate strength, either at or in the vicinity of their own villages. These works were to be supervised by the revenue officers of various grades, aiming at improving the communications or sources of irrigation, and at the same time suitable for unskilled labour. In Madras city, the Public Works Department (PWD) had been the main governmental department responsible for constructing large irrigation works and railways in non-famine periods, and labour was recruited according to ability to perform tasks. During famine periods, the public works were divided into civil works and PWD works. These were intended to provide employment to the famine-stricken poor. The revenue or civil works were to be run on the principle that none who applied for employment could be refused. The wages were also supposed to ensure that people were paid enough for subsistence but no more. The works were organized and supervised by district collectors and other revenue officials who were supposed to prevent starvation. Employment on PWD works during famine periods was offered at rates that assumed that the worker had the necessary strength to put in a full day's work. PWD works were generally supervised by the district engineer of the PWD.[88]

During the famine, the able-bodied poor were sent to PWD works, where all who could perform the task were employed. For those who were too weak to perform the full PWD task, employment on civil works was intended to provide a subsistence wage without running the danger of demoralizing the famine-stricken poor through the provision of gratuitous relief. Classifying a work as a civil famine relief work was intended to ensure that even those who were weakened by hunger would be assured employment. The Madras government held the view that it was desirable to avoid as much as possible the collection of large bodies of labourers, as the rate of wages should not exceed that required for their maintenance. It was felt desirable that the workers were not required to leave their homes.[89]

The main works undertaken in Madras city were for improving the water supply and for constructing a water *kulam*. The Madras government pursued a policy of setting up small relief works under the management of district collectors. The English East India Company state intervened only in the form of employment on useful works and the distribution of gratuitous relief. Of these forms of relief, the latter was increasingly seen as undesirable and as encouraging dependency and demoralization, reflecting ideologies of political economy that shaped the Poor Law introduced in England in the early nineteenth century.[90] However, the extent to which these principles were applied in Madras varied greatly.

THE GREAT FAMINE OF 1875-8 IN TAMIL COUNTRY AND THE APPOINTMENT OF THE FAMINE ENQUIRY COMMISSION AND ITS REPORT, 1880

The transfer of power from the Company to the Crown occurred in 1858. There was a drought in 1876 and it was experienced throughout Tamil country. It continued for the next two years.

Since it was the responsibility of the Crown to mitigate droughts and famines, a Famine Enquiry Commission was established in 1880. Richard Strachey was appointed the first Famine Commissioner. This Commission examined the system of administration, land tenures, irrigation and agricultural improvement. The report of the Famine Commission emphasized the duty of the administration to take measures to prevent the famines as the government stood in the place of the landlord to the agriculturalists. The report specifically recommended the diffusion and application of a scientific knowledge and the provision of a scientific knowledge of agriculture. The report advocated a more systematic collection of meteorological data in order to predict famines.[91] As a follow-up, agricultural departments were created. John Augustus Voelcker was appointed by the government to investigate avenues for further agricultural improvement and to evaluate the progress of the implementation of the report of the Famine Commission. Voelcker, in his report on the improvement of agriculture in 1893, declared that he believed that there was little or nothing that could be improved and that what was necessary was better facilities.[92]

According to Ambirajan the economic-historian, colonialism had played a role in shaping the famine policy.[93] The famines of 1888-9 and 1897 had been called 'Great Famines'. According to the report of the Famine Commission in 1901, the country experienced 12 famines and four severe food shortages between 1765 and 1858. This information of the British is called into question because we have provided statistics of a greater number of famines for the earlier years, particularly culled from the various official and authentic records that go against this famine commission report. Perhaps the Famine Commission suppressed the information deliberately or it did not collect the precise and correct information. However, the famine report admitted that famine and drought calamities were caused by natural phenomenon and human agencies had no control over it.[94] It was also believed that the British Raj in India had to be governed for revenue and not expenditure to be incurred. The Malthusian Theory was also applied and it had been added by the British whereby famine was regarded as a natural check to overpopulation, relieving the state and government from the responsibility of expenditure on relief.

THE FAMINES IN PONDICHERRY IN 1876, 1878, 1885, 1892 AND THE RELIEF WORKS BY THE FRENCH COLONIAL GOVERNMENT

The Great Famine also affected the region of Pondicherry and it was reported to France, seeking financial aid. On 6 February 1877, Rear Admiral Roussin, Deputy Secretary of State, telegrammed from Paris confirming a vote by the assembly to approve a credit of 1,00,000 francs for the famine relief in Pondicherry.[95] This action from Paris was remarkable owing to the extreme urgency of the situation in Pondicherry. Pondicherry had been largely self-sufficient prior to this great famine. It had been rarely necessary for the colonial administration to request direct aid from Paris, as the ministry noted with due appreciation.

In October 1877, Governor Adolph Trillard asked permission to publish a public subscription for aid against the famine. Vice Admiral Albert Auguste Gicquel des Touches approved the request in a telegram from Paris on 27 October, saying that the request for donation would be distributed as an insert in *Le Moniteur de la Flotte*, a publication whose intended audience included every person served by the French fleet.[96]

It is found in the records that many natives of Pondicherry extended considerable assistance during the Great Famines in 1877-8. Governor Trillard, in his communication dated 28 December 1877, commented regarding the favourable change in monsoon conditions and the influence of the north-east winds which had already exerted an effect on public health. Admiral Pothuau also expressed his gratitude to those who showed their sympathy during the famine crisis that traversed the French colony. The governor in his dispatch to Paris requested that several honourable natives like Sandira Pillai, S. Subburaya Pillai, Arunagiri Chetti and Chinna Krishna Chetti should be recognized for their tireless efforts during the famine in 1877. Admiral Pothuau agreed to giving each man a gold medal and assuring the governor that he would consider more commendations. He closed his communication with an expression of personal gratitude and that the French colony was finally released from this long famine crisis. He expressed his gratitude to them for their devotion and admirable charity and for the measures which they had taken.[97] A Central Collection Committee was established with both French and Indian members including the Tamil Councillor General, Mariadas Duraisami Pillai, and K. Lakshmana Sami Chettiar, for mobilizing funds for the famine relief.

The famine relief aid in the form of government guaranteed loans and ex-gratia payments had been found common in Pondicherry since the famine of 1830. The Great Famine of 1876-7 elicited widespread sympathy in France and it prompted the first outpouring of private sector relief subscriptions. On 28 December 1877 in the Ordinary Session of the Colonial Council, Ponnu Thambi Pillai spoke eloquently at length, and he said that in the face of the dolorous circumstances that had come to India in the name of his constituents, to publicly honour the great solicitude of the government and the remarkable devotion of the generous souls who had done such exceptional famine relief work. The merchants, moved by noble sentiment, were united in their efforts to assure food provisions to Pondicherry. Indeed, after encouraging a great many donations in both India and France, the aid raised by the representatives of this chamber as well as those placed in local government, a lot of good had been achieved. The Indian population was profoundly moved by the French interest and generous assistance and this public expression of the Indians' sincere gratitude was the only thanks that was in their power to give.[98]

Ponnu Thambi Pillai in his speech noted the combined efforts of government and the private sector, including the contributions of the Catholic Church and the foreign mission society of Paris besides many honourable residents of Pondicherry. He was also keen to mention the assistance of the British Viceroy of India, who agreed to export grain from English ports to Pondicherry and Karaikal for the duration of the famine, noting with pleasure the humility of the French governor in asking for foreign aid.[99]

Adolph Trillard, the French governor, in his comments in the opening session, noted other factors in mitigating the famine crisis. He spoke of charitable alms of François Jean Marie Laouenan, the bishop of Pondicherry, and the missionaries, whose distribution of abundant alms had lightened the load of the French administration to a great degree. Trillard also noted a lessened severity of the famine in Karaikal.[100]

In spite of the measures taken by the French and British colonial governments in Pondicherry and Madras, large-scale migration of the coolie labourers to overseas developed during the famine years. People signed labour contracts for a period of five years, introduced since 1841, for a sum of Rs. 7 per annum. As many as 87,083 labourers had migrated from Pondicherry to Mauritius between 1849 and 1882; 32,000 coolies sailed to French Guyana; and 25,509 to Guadeloupe between 1851 and 1879. Similarly, 1,19,815 coolies migrated from Madras to Mauritius under the British during 1850–90; 27,72,904 labourers sailed to Sri Lanka between 1839 and 1883 from Tamil ports; 18,634 coolies migrated to Penang in Malaya; and 12,973 labourers went to Singapore between 1850 and 1862.[101] Such large-scale migration of 30,68,918 labourers happened during the 50 years between 1841 and 1890, particularly during famine and drought conditions.

CONCLUSION

The connection between climate, famine and drought continues to be a paradigm in historical research. Decrease from the normal precipitation over an area caused meteorological drought. Reduction in groundwater led to hydrological drought. Inadequate rainfall had led to agricultural drought. However, hunger and scarcity of food had been linked to outbreak of famines. In turn, it had also been connected to bad harvests in the same chain. Thus, the link between climate and famines in the past had been established. Price escalation during famines has been a common phenomenon. People who starved entered into slavery. Individuals and *sabhas* took steps to help people, while the Pallava, Pandya, Chola and Vijayanagara governments had not provided adequate facilities to alleviate the distress of the people. The Portuguese traded in slaves during famine at Nagapattinam and the Dutch had also indulged in slave trade during famines at Pulicat and Nagapattinam. People had offered themselves as slaves to escape starvation and death. The French missionaries in Pondicherry converted several Hindus during famines by providing help to the people. In Madras, the English took efforts to control food-grain scarcity. They provided employment to the people and paid wages in cash. In Pondicherry, the French colonial government took efforts to help people during famines through grant of reliefs and the missionaries and several generous individuals came forward to offer donations and also gave provisions to the needy.

The data collected from the inscriptional records besides Portuguese, Dutch, German, French and English documents regarding the famine years have been carefully examined. While comparing the data from the ninth to the nineteenth century, we find that famines revisited for particular years of different centuries. This proves that the rainfall pattern had not changed with absence of rains in Tamil country to a larger extent. It is easy to trace and find the changing famine years for the long term and the repetition of the famines in the same years in Tamil calendar of 60 years for different centuries is noted. Hence, we may affirm that nature functioned in its own way but man was not able to foresee and forestall the famines. With agricultural economic progress, famines also disappeared.

NOTES

1. C. Minakshi, *Administration and Social Life under the Pallavas*, Madras: University of Madras, 1958, p. 114.
2. Ibid., pp. 114-18.
3. C.B. Namasivaya Mudaliar, *Irayanar Ahapporul: Traditional Commentary based on Nakkirar*, Madras, Paari Nilayam, 1943, p. 29.
4. M. Rajamanickam, *Pallavar Varalaru*, Madras: South India Saiva Siddhanta Pathippu Kazhagam, 1976, p. 245.
5. *Annual Report on Epigraphy* [hereafter *ARE*], Including Indian and South Indian epigraphy from 1881 to 1945, for the years 1887-1981, Madras: Government of India, 404 of 1961-2.
6. Archaeological Survey of India, *South Indian Inscriptions* [henceforth *SII*], Madras & New Delhi: Archaeological Survey of India, 1890 to 1990, vol. II, no. 98, p. 508, verse 14.
7. *Epigraphia Indica* [hereafter *EI*], Calcutta & Delhi, XXII vols., Government of India, 1892-1978, vol. XVIII, no. 2, verse 10.
8. *SII*, vol. II, no. 205, verse 57, p. 396.
9. *ARE*, 404 of 1961-2.

10. *ARE*, 521 of 1920; *ARE*, 1921, part II, para 35.
11. *ARE*, 191 of 1925; 258 of 1925.
12. *ARE*, 274 of 1909; 279 of 1909.
13. *ARE*, 86 of 1911.
14. *ARE*, 560 of 1902.
15. *ARE*, 114 of 1927-8.
16. *ARE*, 1910, part II, para 29.
17. *ARE*, 225 of 1927.
18. *SII*, vol. III, p. 191; *ARE*, 5 of 1899; K.A. Nilakanta Sastri, *The Colas*, reprint, Madras: University of Madras, 1975, p. 562.
19. K.A. Nilakanta Sastri, *The Colas*, p. 400.
20. *ARE*, 560 of 1902.
21. *ARE*, 86 of 1911.
22. *ARE*, 53 of 1899.
23. *ARE*, 29 of 1911.
24. *ARE*, 239 of 1906.
25. *SII*, vol. VIII, no. 303; *ARE*, 30 of 1903.
26. *ARE*, 274 of 1909-10; 279 of 1909-10.
27. *ARE*, 206 of 1903; 207 of 1903.
28. K. A. Nilakanta Sastri, *Studies in Cola History and Administration*, Madras: University of Madras, 1932, p. 245.
29. *ARE*, 5 of 1899, part II, para 53.
30. *ARE*, 520 of 1921, part II, para 35.
31. *ARE*, 521 of 1920; *ARE*, 1921 part II, para 35.
32. *ARE*, 560 of 1902; K.A. Nilakanta Sastri, *Studies in Cola History*, pp. 142-3.
33. *ARE*, 649 of 1902.
34. *ARE*, 397 of 1913-14.
35. *ARE*, 279 of 1909-10.
36. *ARE*, 1899, para 53; *ARE*, 397 of 1913-14.
37. *ARE*, 448 of 1937-8.
38. *ARE*, 291 of 1950.
39. P. S.R. Appa Rao, ed., *Sri Kalakhasti Mahatmyam*, Hyderabad: Andhra Pradesh Sahitya Academy, 1968, canto III, verse 148-56, p. 104.
40. Ibid., verse 150, 152, 153, 156.
41. Gaspar Correa, *Lendas da India*, 4 vols., reprint, Porto, 1975, vol. IV, pp. 131-2.
42. S. Jeyaseela Stephen, *Subaltern Lives: Tamil History and Society from Medieval to Early Modern Age*, Delhi: Kalpaz Publications, 2019, pp. 145, 157.

43. S, Jeyaseela Stephen, *Kaalaniya Thodakka Kaalam, 1500-1800*, Chennai: NCBH, 2018, pp. 154-5.
44. W. Ph. Coolhas, ed., *Generale Missiven van de Gouverneurs-Generaal en Raden Aan Heren XVII der Verenigde Oostindische Compagnie*, 7 vols., The Hague, 1960-84, vol. III, pp. 355-6 (26 January 1661).
45. Joseph Bertrand, *La Mission du Madure d'apres des Documents Inedits*, 4 vols., Paris: Imprimerie de Poussielgue, 1847-54, vol. III, pp. 129-30.
46. Ibid.
47. Sophia Pieters, *Memoir of Hendrick Zwaden 1697*, Colombo: Government Press, 1911, pp. 6-7.
48. Nationaal Archief [hereafter NA], Den Haag, MS Verenigde Oost-Indische Compagnie [hereafter VOC], 1082, fls. 272-4.
49. Tapan Raychaudhuri, *Jan Company in Coromandel, 1605-90: A Study in the Interrelations of European Commerce and Traditional Economies*, The Hague: Koninklijk Instituut voor Taal-Land-en Volkenkunde, 1962, p. 166.
50. Ibid.
51. NA, VOC 1156, fl. 279; VOC 1147, fl. 574.
52. NA, VOC 876, fl. 281.
53. Yvonne Robert Gaebelé, *Une Parisienne aux Indes au XVIIe Siècle: Madame François Martin*, Pondichéry: L'Imprimerie du Gouvernement, 1937, pp. 68-9.
54. Henri Castonnet Des Fosses, *L'Inde Française avant Dupleix*, Paris: Germain & Grassin, 1887, p. 155.
55. Edmond Gaudart and Alfred Martineau, eds., *Procès-Verbaux des Deliberations du Conseil Superieur de Pondichery du 1 er Fevrier 1701 au 31 Decembre 1739*, 3 vols., Pondicherry: L'Imprimerie du Gouvernement, 1913-15, vol. II, p. 207.
56. Edmond Gaudart and Alfred Martineau, eds., *Correspondance du Conseil Supérieur de Pondichéry avec le Conseil de Chandernagor du 30 Septembre 1728 au 10 Fevrier 1757*, 3 vols., Paris: Pondicherry: Societe de l'Histoire del'Inde Francaise, 1915-19, vol. II, p. 10.
57. Frederick Price, H. Dodwell, and V. Rangachari, trans & ed., *The Private Diary of Ananda Ranga Pillai* [hereafter *ARP Diary*], 12 vols., reprint, Delhi: Asian Educational Services, 1980, vol. XI, pp. 446-7.
58. Pierre M. la Mazière, *Lally Tollendal*, Paris: Plon, 1931, p. 65.
59. Archives Nationales [hereafter AN], Paris: MSS *Compagnie des Indes et Inde française*, Colonies Série C^2, vol. 196, fol. 2.
60. Edmond Gaudart and Martineau, *Correspondance du Conseil Supérieur*, vol. II, p. 10.

61. *ARP Diary*, vol. V, pp. 395-6.
62. Ibid.
63. Ibid., p. 228.
64. Ibid., pp. 273-4.
65. Pierre M. la Mazière, *Lally Tollendal*, p. 165.
66. Gobalakichenane, *Irandaam Veera Nayakkar Naatkurippu, 1778-92*, Madras: Nattamil Pathipagam, 1992, pp. 174-5.
67. Ibid., p. 260.
68. Ibid.
69. Lettre du Père Claude Barbier, Jésuite au Père de la même Compagnie, Pinneipundi, Mission de Carnate, 15 January, in *Lettres Edifiantes et Curieuses Ecrites des Missions Etrangeres par Quelques Missionaries de la Compagnie de Jesus* [hereafter *LEC*], 28 vols., Paris: Chez N.E. Sens, Chez J. Varnavel, 1780-94, tome 13, pp. 296-7.
70. Archives de la Societe des Missions Etrangeres de Paris [hereafter AMEP], Paris, *Correspondance Pondichery-Paris*, vol. 993, Lettre de Louis Mathon, prêtre, 21 December 1774; Extrait d'une lettre de Père Calmette au Père de Tournemine, Vencataguiry, 16 September 1737, in *LEC*, tome 14, p. 10; Lettre du Père Poissevin de 4 et 16 December 1743 à Madame Hyacinthe, Crichnapouram, in *LEC*, tome 14, p. 173; Henri Castonnet Des Fosses, L'Inde Française, p. 155.
71. Extrait d'une lettre du Père Poissevin au Père d'Irlande, Chandernagore, 11 January 1749, in *LEC*, tome 14, pp. 249-51.
72. Lettre du Père X de Saint-Estevan à M. Le Comte, Pondichéry, 7 December 1754, in *LEC*, tome VIII; Adrian Launay, *Histoire des Missions de l'Inde: Pondichery, Maissour, Coimbatour*, 5 vols., Paris: Societe des Missions Etrangeres, 1895-8, vol. I, pp. xxxviii-xxxix.
73. AMEP, *Correspondance Pondichery-Paris*, vol. 998, p. 776; Lettre à M. Tesson, Karikal, 10 July 1833; AMEP, *Correspondance Pondichery-Paris*, vol. 998, p. 776; Lettre de Pierre Brigot à M.Tesson, 10 July 1833; Adrian Launay, *Histoire des Missions*, vol. I, pp. 293-4.
74. Jacques Weber, *Les Etablissements francais en Inde au XIX e Siecle (1816-1914)*, 5 vols., Paris: Librairie de l'Inde, 1988, vol. I, p. 576.
75. Adrian Launay, *Histoire des Missions*, vol. IV, p. 25.
76. AMEP, *Correspondance Pondichery-Paris*, vol. 1004, Lettre de M. Fourcade à M. Ligeon, Nangatur, 10 October 1874; AMEP, *Correspondance Pondichery-Paris*, vol. 1004, Lettre de Mgr. Laouenan à MM. Les membres du Conseil de la Propagation de la Foi, 4 November 1871; Adrian Launay, *Histoire des Missions*, vol. III, pp. 463-64; vol. IV, p. 25.
77. AMEP, *Correspondance Pondichery-Paris*, vol. 1004, Lettre de Mgr. Laouenan

à MM. Les membrȩs du Conseil de la Propagation de la Foi, Pondichéry, 18 December 1869.

78. Jacques Weber, *Les Etablissements francais*, vol. I, p. 576.
79. AMEP, *Correspondance Pondichery-Paris*, vol. 1006, Lettre de Mgr. Chevalier aux Missionnaires, Bangalore, 6 February 1878; AMEP, *Correspondance Pondichery-Paris*, vol. 1004, Lettre de Mgr. Laouenan à Mgr. Hugarin, Evêque de Bayeux, Pondichéry, 28 June 1878; AMEP, *Correspondance Pondichery-Paris*, vol. 1004, Lettre circulaire de Mgr. Laouenan aux Missionnaires de son Vicariat, Pondichéry, April 1877; AMEP, *Correspondance Pondichery-Paris*, vol. 1004, Lettre de Mgr. Laouenan aux Missionnaires de son Vicariat, Pondichéry, 31 December 1877; AMEP, *Correspondance Pondichery-Paris*, vol. 1004, Lettre de M. Fourcade à Mgr. Laouenan du 17 April 1877 et 9 June 1877, Alladhy; AMEP, *Correspondance Pondichery-Paris*, vol. 1004, Lettre circulaire de Mgr. Laouenan aux Missionnaires, Pondichéry, 8 July 1878.
80. Adrian Launay, *Histoire des Missions*, vol. III, p. 31.
81. R.A. Dalyell, *Memorandum on the Madras Famine of 1866*, Madras: Government Press, 1867, pp. 18-19; W. Francis, *Madras District Gazetteers: Madurai*, Madras: Government Press, 1906, pp. 161-2.
82. W. Francis, *Madras District Gazetteers: South Arcot*, Madras: Government Press, 1906, p. 179.
83. R.A. Dalyell, *Memorandum*, p. 21.
84. Ibid, p. 23.
85. Ibid, p. 29.
86. Ibid, p. 26.
87. British Library [hereafter BL], London, Oriental and India Office Collection [hereafter OIOC], V/4/Session 1867/vol. 52, *Copies of Papers Relating to the Famine in Madras Presidency*, no. 35, Proceedings of the Madras Government, Revenue Department, 25 July 1866, enclosed in GoM to SoS, dated 11 August 1866.
88. Ibid.
89. R.A. Dalyell, *Memorandum*, p. 26.
90. G.E. Royer, *An Economic History of the English Poor Law, 1750-1850*, Cambridge: Cambridge University Press, 1990, p. 84.
91. BL, *Report of the Indian Famine Commission*, Part I: *Famine Relief*, London: Madras: Government Press, 1880, p. 126.
92. BL, John Augustus Voelecker, *Report on the Improvement of Agriculture*, London: Madras: Government Press, 1893, p. 45.
93. S. Ambirajan, *Classical Political Economy and British Policy in India*, Cambridge: Cambridge University Press, 1978, p. 214.

94. BL, *Report of the Indian Famine Commission*, London: Madras: Government Press, 1880, p. 66.
95. *Bulletin Officiel des Etablissements francais de l'Inde, 1877*, Pondichery: L'Imprimerie du Gouvernement, 1876-1910, Arrête No. 76, 6 Fevrier 1877, p. 12.
96. Ibid., *Bulletin Officiel*, Arrête No. 376, 27 October 1877.
97. Ibid., Arrête No.175, 11 April 1878.
98. National Archives of India, [henceforth NAIP], *Proces-verbaux Conseil Colonial de Pondichery de 1877*, Séance du 28 Decembre 1877, Session Ordinaire, Puducherry: Regional Record Centre, pp. 4-5.
99. Ibid.
100. Ibid.
101. S. Jeyaseela Stephen, *Goodbye to Tamil Motherland: The Rise of Labour Migration Overseas and the Society, 1729-1890*, Delhi: GenNext Publication, 2019, pp. 145-9; S. Jeyaseela Stephen, *Kaalaniya Valarchi Kaalam, Pulampeyarnthavargalin Vaazhai*, Chennai: NCBH, 2018, pp. 157-64.

CHAPTER 4

The Storms and Cyclones of Tamil Littoral and the Europeans, Seventeenth-Nineteenth Centuries

Sailing in the Bay of Bengal and knowledge of the winds, currents, storms and cyclones had been well known to Tamils through tradition, after keen observance and experience being handed over through generations. The sailors knew the various types of winds that blew in different directions. Any violent wind accompanied with rain was termed as storm in the medieval period. Those that were lasting and belonged to a destructive type were known as cyclones or cyclonic storms. The fishermen found a way to balance the risks associated with living near the sea. Thiruvalluvar in his *Thirukkural* mentioned that 'Even the wide sea will be depleted if the clouds do not give of themselves'.[1] This couplet ties the wealth of the ocean to the inevitable regularity of monsoon. Madurai Marudhan Ilanaaganaar, the poet mentioned that the ships had split the water while sailing as if the earth has turned down. The ocean had the smell of fish. The waves had been tossed by swift winds that continued to blow day and night. The helmsmen steered the sailing vessel guided by the bright lamps of the oil lit in the top of the mansions so as to reach safely the shores heaped with sand. Thus *Aganaanuru*, the Sangam literary text mentions how the Tamils sailed across the oceans.[2]

Tamil sailors who accompanied the merchants in conducting overseas trade knew the winds and the monsoon pattern and so they sailed across the seas in the months of May-June from the Coromandel Coast to Thailand and other countries in Southeast Asia. They also safely returned by November through the winds that blew. The eighth-ninth centuries Takua-pa inscription in Tamil (now preserved in the Nakhon Si Thammarat Museum, Thailand) records a *kulam* called Sri Naranam that was dug by Nangarajyan and it was placed under the protection of members of *manigramam* (the trade guild).[3] Thus, storage of water had been given priority by Tamil merchants and travellers who visited foreign lands. A precise knowledge of winds, tides, and currents around the coastline in the Tamil region was essential to minimizing sailing time and thereby maximizing profit in trade and commerce.

The early Europeans who came, traded and established their settlements on the Tamil coast and began to understand the seasons, winds, tides, etc., of the Bay of Bengal. Over a period of time during their stay, they precisely collected relevant information that was necessary for smooth sailing. Tome Pires, a Portuguese traveller, noticed in AD 1511 that it was the Coromandel Coast at the high-rising southern hills and plateau which prevented the south-west and west winds to blow in the kingdom of Vijayanagara. He said that the fresh winds that had been blowing while they were coming from Sri Lanka to the coast of India, ceased blowing when the sailor reached the Coromandel Coast.[4]

THE STORMS AND CYCLONES IN MADRAS IN AD 1640, 1662, 1668, 1684, 1687, 1695 AND THE DAMAGES CAUSED TO BOATS, SHIPS AND BUILDINGS

The Tamil coast had been damaged by heavy storms and cyclones and we have little written records prior to the coming of the Europeans. In the age of expanding commerce, the European trading companies began to write down the details of the storms and cyclones that havocked the region. According to the English Company records, a severe storm struck Armagon (Durgarajapattinam) on 12 March 1640 at such a time of the year that even the oldest man living among the local people had not seen before.[5] The local nayak ruler arrived and treated the English crew kindly and directed the salvage operations. Mr Carter, the captain of the ship called *Eagle*, sent a report to Mr Cogan, the English Company official, at Madras from Armagon. Thereafter, Henry Green Hill the English Company servant in Madras was sent down with relief measures.[6] The records maintained by the English continue to furnish very detailed information on the weather and climate. It has been mentioned that the English Company lost two of its ships in Madras during the heavy storm of 12 March 1640.

In February 1662, another storm occurred in Madras and the Fort St. George Council said that the ferocity of that particular storm at that time of the year had never been seen or known in the knowledge of any man alive then. The weather was terrible and there were shipwrecks at Madras and Porto Novo and that this was a sad fate.[7] Another storm on 28 February 1662 had elicited the same response. Nine ships were lost in Madras, while 21 more ships were also lost on the Coromandel Coast. The storm also damaged Porto Novo.

A dreadful cyclone struck Madras on 22 November 1668. The tempest and rain were so violent that nothing could withstand its onslaught—houses and doors collapsed, trees broke down, and none were left standing either in a garden or elsewhere. Nor could the English Company protect people, books or papers from the violent wind and the rain. It was expected that around 3,000 *pagodas* would be needed to be spent at first thought.[8]

An intense storm on 3 November 1684 destroyed several boats. The buildings in Fort St. George were severely damaged. Houses in the native settlement of Chennapatnam were devastated. The estimated amount of loss was said to be 20,000 *pagodas.*[9] Cyclones during 4-8 October 1687 wrecked two ships of the English Company and two country boats of Fort St. George 30 boats were destroyed.[10]

Likewise, in the cyclone of 22 November 1695 in Madras, many boats laden with cargoes were lost. Government buildings were damaged; roofs of houses in native quarters were blown away; part of the fishing village of Chennapatnam was washed away; and numerous trees were uprooted.[11]

THE CYCLONE AT SANTHOME IN MYLAPORE IN AD 1674

A terrifying cyclone over Santhome in Mylapore on 21 April 1674 resulted in death and destruction. Waves had removed the eastern wall and countless people died and the dead were buried on the beach. It was reported in Madras that the workmen sent by the English recovered what they could from the wreck. Some had died on sea and their bodies had washed ashore. It was mentioned that the storms that continued unabated from 21-23 April caused much loss of life. A French ship which had halted at Mylapore was lost, while two Dutch ships in Pulicat were also lost.[12]

THE VIOLENCE OF THE STORMS IN MADRAS IN AD 1717, 1721, 1729, 1730, 1746, 1749 AND 1752

It has been mentioned in records that four vessels were damaged in the roadstead during the storms on 7-8 April 1717. Several vessels were found missing from Santhome in Mylapore.[13] The storms on 13-14 November 1721 were so severe that the countryside was flooded. Bridges were demolished and the fishermen area of Chennapatnam was washed away. Three large ships of the English Company were lost.[14] When a cyclone struck Madras on 30 October 1729, the buildings in Fort St. George were damaged because of continuous rain, as was the bridge at Triplicane. Many trees in the island near River Kuvam were uprooted. Roads were washed away by the waters.[15] The English settlement of Madras was affected again on 28 November 1730 because a violent storm caused destruction to houses, walls, bridges, including the Protestant missionary house.

On 2 October 1746, an *ouragan* (hurricane) that lashed the searoads between Pondicherry and Madras destroyed a squadron that was on its way to blockade the English at Fort St. George in Madras; this squadron was under the command of Bertrand-François Mahé de La Bourdonnais. The ship *Duc d'Orleans* sank with almost all hands. Likewise, the *L'Achille*, the *Bourbon*, and the *Neptune* all foundered, losing large parts of their artillery and much of their cargo, as well. Nearly 1,200 people had perished in the storm.[16] On 14 October 1746, a violent storm blew up at Madras, sinking four French ships. A Dutch ship was stranded near Madras. Many dead bodies were later found on the beach.[17] The English Company records mentioned that on 1-2 October 1746 great damage was caused because of floods. Two French warships were lost and four other ships were blown out to sea and dismasted. Three large ships of the English Company submerged during the storm of 14 October 1746, resulting in the death of 1200 men.[18]

It has been mentioned that on 24-5 April 1749 in Madras, an intense north wind in the first part of the storm blew up to midnight, felling trees and causing some damage to rooftops, but the Lutheran missionaries recorded that much stronger south winds grievously damaged houses and even blew down the strongest trees.[19]

On 31 October and 1 November 1752, a cyclone struck Madras and damaged the fort and other government buildings.[20] The gates of Chennapatnam town were blown away. According to the Lutheran missionaries, there was a severe thunderstorm accompanied by lightning since the midnight of 31 October 1752. After one-and-a-half hours, it became stronger when south winds set in; it continued thus until 6 a.m. in the morning, and was followed by a violent storm on 1 November 1752 over Madras and Pulicat, with north winds. Buildings were devastated, and many lives were lost. *Colchester*, an English ship, was swept southwards to Cuddalore with broken masts. Heavy storm winds continued to blow along the coastal stretch between Pulicat and Sadurangapattinam.[21]

THE STORMS AND CYCLONES IN MADRAS IN AD 1763, 1768, 1773, 1778, 1780, 1782, 1787, 1791, 1797 AND THE SUFFERINGS OF THE PEOPLE

It is reported that a tempest lasting 14 hours struck Madras with tremendous force in AD 1763. Every ship at sea was destroyed, with almost all hands lost. A cyclone from 20-22 October 1763 destroyed several vessels.[22] English Company officials reported that buildings were unroofed and the hospital was flooded because of heavy rains on 30 October 1768. Cattle died in large numbers. The English Company came to the assistance of the suffering inhabitants of Madras. According to the report of the Lutheran missionaries, the storm had begun in the evening of 30 October 1768 and around 250-60 lives were lost in the black town of Madras. The wind began as a north-easterly, turned east during the evening and streng-thened at midnight. It finally turned south-east and then south-west, when it was at its most furious.[23]

A cyclone on 21 October 1773 damaged all the ships anchored in Madras, and 100 sailing vessels were lost.[24] In the end of November 1778, a fierce storm farther north caused considerable damage. A large English ship laden with 12,000 sacks of rice sank with its crew at Madras while another was stranded at Pulicat.[25]

There was a storm in Madras in 1780.[26] The English Company records state that on 14-15 October 1782, a fierce cyclone had damaged 10 ships and 195 country vessels.[27] It also destroyed every indigenous craft loaded with rice.[28]

The Lutheran missionary records mention a storm during 17-21 May 1787, which was so severe that the sea rose 14 ft high at Madras, leading to massive coastal inundation. Its effects were felt as far as Vizagapattinam, and the damage was severe with loss of thousands of lives. The sea inundated the land for several miles and destroyed a number of villages. All houses were destroyed and left unfit for habitation. Most of the trees were either denuded of their branches or blown away.[29] The annual report of Jacob Klein, the Protestant missionary, thus described the effects of the storm of May 1787 in minute detail.[30] Most cyclones were accompanied not only by strong winds that wreaked havoc on housing, but also caused severe river-flooding.

All rivers were flooded in Madras on 26 December 1789 and 13-14 November 1791. The Marmalong bridge on River Adyar near Saidapettai had breached its banks. The details of the destruction resulting from a cyclone that hit Madras on 27 October 1797 has not been found in English records.[31]

THE MADRAS WEATHER DIARY OF THE PROTESTANT MISSION, AD 1789-91

A daily weather diary was kept at the Protestant mission in Madras from 1 February 1789 to 31 December 1791. The name of the missionary who maintained it is unknown. This manuscript has 77 sheets of tabulated data and weather comments. There are no monthly or annual summaries, but these are occasional notes on weather observation and cloud cover classification, such as the sky was more clear and less clouds; clear sky and clouds were equal; more clouds than clear sky; the whole sky was covered with thin clouds; there were some thick clouds; the majority of the sky was covered with thick clouds; and the whole sky was covered with thick black clouds. The observer described cloud cover on the basis of his observations made at midday. The German missionary in Madras had given details of day-to-day observation of air pressure, temperature and wind direction in 1789 in tabular column.[32] The annual meteorological data had also been prepared in tabular form by the same missionary in 1790.[33]

The wind defined was called as north-east and south-west monsoons. The north-east monsoon, taking the mean of the years, usually set in on 19 October and ended around 2 March, with heavy rainfall sometimes for several days together from its commencement to the middle of December, attended at times by gales of wind. Thereafter, till the close of this monsoon, the air was generally clear and cool, and the weather pleasant. This agreed closely with the practice of lowering the flag-staff of the Fort at Madras on 15 October, when the stormy season was considered to commence. It was hoisted again on 15 December, after which it was supposed ships might approach the coast with safety.[34]

THE LUTHERAN MISSIONARY ACCOUNTS OF THE STORMS AND CYCLONES IN TRANQUEBAR IN AD 1710, 1723, 1749, 1754, 1778, 1783, 1787, 1789 AND 1794

The Danish-Halle missionaries recorded the anomalous climatic events of the ports on the Tamil coast where they lived. They mentioned details of drought, heavy rain, violent winds and exceptional cold, as well as climate-related events such as river and coastal flooding, heavy seas, crop failure, food prices and famine. These are useful for purposes of reconstructing the history of the climate of the Tamil coast and hinterland. The Lutheran missionaries in Tranquebar wrote that on 10 November 1710 trees had been blown over and defoliated by a cyclone. The cyclone had also caused massive river-floods in Tranquebar and surrounding villages. 20 houses had been destroyed by a tidal wave near the town wall.

Again, on 23 December 1723, Tranquebar and the Porto Novo coasts were struck by a cyclone which caused considerable damage to housing and property. The missionaries reported that Nagapattinam, Tranquebar and Karaikal areas were severely affected during the period 22-25 April 1749 and on 8 November 1754 when a fierce storm gathered from the north-east between 6-10 a.m. and was then followed by south-east winds which blew strongly till the evening. Houses and gardens at Nagapattinam were severely damaged. Streets in the Dutch settlement were flooded. The storm had an impact as far as Tiruvarur and Thanjavur in the interior.[35]

The missionaries of Tranquebar in their accounts noted weather observations for the entire month of December 1776. They said that in the end of November 1778, a storm hit Tranquebar from the south-west. All the streets in the town became flowing streams. On 2 November 1783, the westerly winds that blew at Tranquebar caused a storm which affected normal life. During 17-21 November 1787 and on 26 December 1789, the north-east wind caused coastal inundation at several places around Tranquebar, including the south side of the town wall. Gardens and fields to the south of Tranquebar were rather badly damaged. From 25-28 October 1794, the north wind blew at Tranquebar, causing heavy rain. On 27 October 1794, the weather was very warm from 9 p.m. onwards but later there was an intense storm. The north-east winds blew from 11 p.m. and later the west wind and storm abated around 10 a.m. on 28 October 1794. Reports of the weather and the storm during the monsoon period of 1794 was communicated to Europe by Christoph Samuel John the Lutheran missionary in his letter written from Tranquebar dated 18 January 1795.[36]

THE STORM AT KARAIKAL IN AD 1754

A small storm that blew at Karaikal in AD 1754 greatly impacted the place. It had been reported that although the seaside of Karaikal was outside the influence of the winds from 8-10 October 1754, houses and trees were levelled and paddy fields were destroyed because of heavy rains.[37]

THE STORMS AND CYCLONES IN CUDDALORE AND PORTO NOVO IN AD 1745, 1749 1754, 1761 AND THE WRECKS

The Lutheran missionaries had reported the weather and climate of Cuddalore since they set up a mission there. According to them, the Coromandel Coast was affected by storms either in the months of May and June or during October and November. It was said that on 23 October 1745, Cuddalore witnessed storms with the wind starting as a westerly and ending as a south-easterly.[38] In November 1745, Cuddalore twice experienced violent winds with much damage to housing.

On 13 April 1749, a severe cyclone hit Cuddalore and 74 guns on a ship were destroyed, and 850 men in boats were blown away from the Cuddalore coast.[39] On 24-5 April 1749, a violent storm blew up at Cuddalore with winds from the north, and from 3 a.m. the direction of the wind changed to the south. Three warships sank, some other ships were wrecked on the seashore, and there was loss of at least 1,000 European lives. There was also heavy damage to houses in Cuddalore.

An ouragan hit Porto Novo in AD 1749 and it is mentioned by Monsieur Floyer the French naval captain in his letter dated 11 May 1749. He primarily made a request for a number of cannons to be brought to Pondicherry. Occurring as it did during a lull in the recently quieted Carnatic wars, this seems a singular request, especially given that the ouragan mentioned had already halted an English expeditionary force near Porto Novo, and Pondicherry itself had not been heavily affected. Floyer's account focused on English losses: the hospital; the magazine at Porto Novo; a ship named *Apollo* with all hands; the 60-gun Pembroke; the 74-gun Namor; and 750 British sailors. The damage to the French navy was limited to one ship and two small boats swamping 60 cannons.[40]

It has been also mentioned that there was storm and heavy rain on 6-8 November 1754. The people at Vandipalayam, Tirupapuliyur and Cuddalore suffered severe damages. The French and English godowns at Porto Novo were washed away. The Dutch factory godowns suffered huge losses because they were under 30 ft water.[41]

On 1-2 January 1761, Cuddalore witnessed storm and heavy rains at 9 p.m. The northerly wind increased, reaching its greatest strength around midnight. The storm continued until 4 a.m. The roofs of most houses in Cuddalore were blown away. The houses of the poor were completely destroyed, and many trees were blown down.[42]

STORMS AND CYCLONES IN PONDICHERRY IN AD 1681 AND 1687

The French ships departed the port of Lorient in February-March and passed through the Cape of Good Hope in April-May and reached Pondicherry in August-October. Similarly, ships sailed from Pondicherry in January-February, passed through the Cape of Good Hope in April-May and reached Lorient in July-August. The records of the French Company officials and servants mention the cyclones that hit Pondicherry. A cyclone was recorded, hitting Pondicherry on 19 November 1681. It is described that the three days' storm had centred over Porto Novo, and destroyed several small boats in the river. Many houses collapsed under the heavy rains and residents migrated to places in the inland, seeking safety and security.[43]

A cyclone hit Pondicherry on 11 April 1687. This resulted in food scarcity, accompanied with the rise of prices. Hence, François Martin, the agent of the French East India Company, arranged for import of rice by boats which was sold in Pondicherry.[44] There was another severe cyclone in Pondicherry on the night of 17 August 1687, which caused considerable havoc. A number of small vessels anchored at the harbour were destroyed, damaged and lost.[45]

THE DESCRIPTION AND DAMAGES CAUSED BY THE THREE CYCLONES IN PONDICHERRY IN AD 1745

On 3 November 1745, a ferocious storm struck the Coromandel Coast. The first report is found in the Tamil diary of Ananda Ranga Pillai, the Chief *dubash* of Dupleix, the French Governor. The French settlement of Pondicherry was devastated by the cyclone after sunset and winds continued to blow very strongly throughout the whole night and caused terrible destruction. Nearly all the trees were uprooted, and many were twisted out of shape. Coconut, mango and other trees in the orchards and gardens were all laid low. The cyclone had disrupted the lives of the people.[46] Ananda Ranga Pillai, thus, mentioned it in great detail, describing the effects of this cyclone. He also added that the town of Pondicherry was inundated, and both people and cattle perished in the rushing flood. Houses, both within the town area and at the outskirts, were laid low and water, in places, stood up to a man's waist. Birds and livestock died in large numbers, their bodies littering the streets for days afterwards. Dead goats were purchased, brought into the town to be butchered and left to dry in houses, but the unrelenting dampness in the following days prevented drying and soon the town was filled with a sickening stench so horrific that people remained in their homes for days, even after the storm had stopped and the flood subsided. Pillai said that because the wind had continued to blow in this way, not a single house in the town escaped.[47] Those who lived in houses erected on the land allotted to them on the banks of the Upparu river suffered the most because the river rose in flood during the night, and houses in all the three streets of this area were swept away. The water rose to a height of the length of the forearm over the ruined buildings. Many people drowned. Several streets in the town were submerged, bringing down several houses. When morning dawned, the wind abated and the rain ceased.[48]

On 10 November 1745, Pillai noted the departure of Governor Dupleix to Mortandi Chavadi on a retreat, because he found the rainy season in Pondicherry a nuisance. On 22 November 1745, the whole night was rocked by violent winds. It was followed by another strong gale which blew for three hours on the night of 23 November 1745. Its force was about one-fortieth of the ouragan which had raged 20 days earlier. This second storm, unlike the first, had been perfectly predictable. On this occasion, everybody was afraid.[49] Pillai opined that the disturbance in the weather had been inevitable due to a convergence of three factors: (1) the new moon; (2) occurred on a Tuesday; and (3) under the influence of the star, Kettai. It is worth underscoring the ill-omen ascribed to this event by Pillai, for he relied on the *shastras* being a Hindu.

On 27 November 1745, there was another storm which blew from 7 p.m. until 9 a.m. the next day. Its strength was either three-fourths or half of that of the first storm (3 November). This storm uprooted the trees that had escaped the violence of the first. The following Sunday, 28 November, the third storm of middling strength again struck the town. Its strength might have been three-quarters of the first storm, but some thought it was only half as strong. Pillai, attributing it to the lesser amount of damage caused, pointed out that all the mischief had occurred during the earlier two storms in the same month.[50] He ended his diary entry with some final words on the uncommon triple cyclone: 'Never before have there been three storms within the same month. What evil times these may be'.[51] According to him, he had never witnessed three storms in a month as that of 3rd, 23rd and 27th November 1745. He reported that 40 people had died and 2,000 houses damaged.[52] He concluded with the words that though everyone was afraid, God had favoured the people with a gentler storm.[53] His account on the storms and flood was more emotional. He wrote of the water level rising to above thigh level, of the hundreds of dead birds littering the streets, of the devastated plantations, the uprooted trees and the stench of rotting animals.

Another report, the second one about the storms of November 1745, comes from a letter dated 11 January 1746, addressed to the directors of the French East India Company in Paris. It stated that 2,000 houses were destroyed and more than 40 people had died. It mentioned that the French colony was struck by two storms, one on the night of 2 November, and the other on 27 of November, which laid waste across the land. However, the storm of 23 November had not been mentioned.[54]

THE IMPACT OF THE CYCLONE OF AD 1745 AT SADURANGAPATTINAM

The cyclone also affected Sadurangapattinam in AD 1745. Many trees were downed in the Dutch settlement and some were half uprooted. Mud houses of the poor were totally destroyed by the wind and rains. Occasional damage to stone houses and walls occurred. The wind blew the strongest between Pondicherry and

Sadurangapattinam. The wind and flooding were more severe in the French settlement of Pondicherry than at Cuddalore. It was reported that all the houses on both sides of a street at Sadurangapattinam had been washed away and 50 people had died.[55]

The Storms and Cyclones in Pondicherry in 1752, 1760, 1761, 1762, 1787, 1795, 1811, 1820, 1845 were severe and the damages to the ship's crew, cannon, landing crafts and pulling down magazines and hospitals were vast.

On 30 October 1752, a cyclone hit Pondicherry and it was followed by incessant rains. Normal life was paralysed. Vast stretches of land in and around Pondicherry were submerged under water. The monsoon season in 1760 then began to grow precarious. Stevens and Cornish, the two English Company naval admirals had prepared to leave the Coromandel Coast during the monsoons and they retired to the island of Sri Lanka, where they could refit the squadron, and shelter it from the storms. They had set sail on the 23 October 1760 and blockaded Pondicherry by sea with five ships along with Captain Haldane. This smaller blockade continued through the end of 1760 when, in December, a powerful cyclone struck Pondicherry.[56] It hit on 30 December 1760, started at 8 a.m. and continued till 10 p.m. Three vessels carrying 1,100 Europeans sank 2 miles south of Pondicherry, leaving only seven survivors.[57] The effect of this storm on the English fleet was most disastrous.[58] The English Company records mention that the *Aquitain* and *Sunderland*, the two with 60 guns and the *Newcastle* with 50 guns, the *Queensborough* with 20 guns, and the *Protector* fire-ship were the ships that were heavily affected. They had foundered and sank offshore, very nearly losing all hands to the tempest. The *Newcastle*, *Queensborough*, and *Protector*, in the general tumult of the storm, were unable to make out the sound of the surf against the strand, and so ran aground 2 miles south of Pondicherry.[59]

On 29-31 October 1761, there was heavy rain. On 1 January 1761, there was a storm at Madras but the centre of the cyclone passed over Pondicherry, causing immense devastation to life and property. On 1 November 1762, a cyclone battered this city. On 20 May 1787, there was a terrible ouragan at Pondicherry which was not usual. In 1795, a cyclone struck Pondicherry and it led to a massive crop loss in the region owing to heavy rains.[60]

In 1811, the ships *The Dorn* and *The Chichester* were completely destroyed during a deadly storm in Pondicherry, together with several other ships and boats. In 1820, a violent tempest in Pondicherry broke out and several landing crafts were lost and there was great damage to livelihoods.[61]

Adolphe Philibert Dubois de Jancigny and Xavier Raymond the two French naval officers in 1845 had recorded a storm that devastated Pondicherry. The English flotilla holding the blockade was broken and dispersed. The sea had burst its limits, the sea-wall, and flooding all the way to the hedge-wall, submerging and washing away the batteries and redoubts of the besieging encampment. The inhabitants of Pondicherry thought they had been delivered, but the tempest ravaged the town, pulling down the hospital and magazines, destroying everything within.[62]

THE DREADFUL STORMS AND CYCLONES IN MADRAS, 1811, 1820, 1827, 1841, 1851, 1859, 1872 AND 1874

Storms and cyclones had appeared in Madras and the details of 1811, 1820, 1827, 1841, 1851, 1872 and 1874 are found in the *Journal of the Asiatic Society of Bengal*, printed in 1877. In 1859, Edward Thornton, the British naval official, said that the English suffered severely from the dreadful storm in Madras. The sea broke over the beach, and overflowed the country, carrying away the batteries and redoubts. Their tents were destroyed and their ammunition rendered useless, while the soldiers, in many instances, abandoned their muskets in their anxiety for personal safety. Many of the native retainers of the camp perished. He also described the strategic effects of the storm, noting that the widespread waters made repositioning artillery impossible and prevented the French from sallying against a British encampment that could only have responded feebly if they had. It is worth noting the common disparity that while many English soldiers abandoned their muskets for safer locations, Thornton said that the Indians were left to drown in the flood.[63]

The reports of the meteorological office in Madras throw considerable light on the various storms that originated from the various ports of the Coromandel Coast from 1842-87.

TABLE 4.1: THE LIST OF STORMS ORIGINATED FROM THE VARIOUS PORTS ON COROMANDEL COAST, 1842-87

Date	Place of Origin
22-24 October 1842	Pondicherry
28 April–1 May 1872	Sadurangapattinam
21 April–5 May 1874	Madras
18-21 May 1879	Madras
15-19 November 1879	Madras
19-21 November 1880	Nagapattinam
10-13 November 1881	Madras
21-24 November 1882	Sadurangapattinam
28-30 November 1882	Cuddalore
15-17 October 1884	Nagapattinam
12-18 December 1884	Nagapattinam
1-3 November 1885	Pulicat
19-21 November 1885	Madras
7-9 December 1886	Madras
22-24 May 1887	Nagapattinam

Source: Henry F. Blanford, *A Practical Guide to the Climates and Weathers of India, Ceylon and Burma, and the Storms of Indian Seas*, London: Macmillan, 1889, pp. 337-41.

THE STORMS AND CYCLONES IN KARAIKAL AND THE SUFFERINGS OF THE PEOPLE, 1818-71

The French records mention various storms and cyclones that hit Karaikal from time to time. We find the description and the damages. On 24 October 1818, a violent storm appeared and heavy rains followed. Many trees were downed in the town; paddy fields were submerged in water; 20 cattle had perished; and people suffered greatly.

On 20 March 1820, a storm and heavy rains pelted Karaikal. The fishing boats and *thonis* on the shore were swept off to the sea and three persons died. The huts of the fishermen were severely damaged. On 6 December 1827, there were strong winds but no deaths of persons and animals were reported. On 17 December 1827, there was a storm with strong winds but no deaths of persons and animals. On 2 December 1828, strong winds blew and subsequently trees fell. There was light rain and it continued for long.[64]

On 2 December 1830, a violent storm blew and floods resulted from the heavy rains. The paddy fields were in deep water and houses were damaged. Boats and *thonis* on the seashore were lost. The French Colonial Government announced reduction in payment of the land tax. A small sum was given as relief. The Governor of Pondicherry sent a report to France mentioning the calamities. On 30 October 1836, a storm tore into Karaikal and the rains that continued affected the paddy fields.

There was a violent storm on 26-27 March 1853. The beach road was swallowed by River Arasalar, as many as 4,000 trees were destroyed and 100 cattle had perished. The huts in the town were virtually destroyed. The French Colonial Government announced relief and gave a small amount to the hut dwellers and those who had lost their cattle.

On 20 November 1856 and 9 December 1856, a fierce storm battered Karaikal and floods hit the region. Farmers were greatly affected because their crops were destroyed. The Colonial Government gave relief to the affected. Heavy rains in 1858 caused floods. The banks of River Arasalar were broken, and paddy fields were havocked. There was water everywhere in the streets. A number of cattle had perished. People took shelter in the churches and temples and food was supplied to them.

On 19 October 1863, there was storm and incessant rain. Boats and *thonis* sank in the sea. The huts of the fishermen near the seashore disappeared. Paddy fields were devastated. The year 1864 saw storms on 19 October, 28 October and 28 November. There were floods, because of which bridges and roads were affected severely. A storm and heavy rainfall pelted the region during the period 25-27 November 1865.

On 6 November and 16 November 1871, a storm and rainfall led to floods. Trees were uprooted fell on the huts. The hospital and the office of the harbour buildings were damaged. Five persons died at Niravai.[65] It may be concluded that storms had occurred in Karaikal 15 times in November. We find occurrence of storms five times in October, five times in December and thrice in March.

JOHN GOLDINGHAM'S SCIENTIFIC STUDY ABOUT THE MOON'S IMPACT ON GALES AND STORMS IN MADRAS, AD 1797-1820

Gales usually struck Madras only during the rainy season, between mid-October and mid-December. The barometer seldom sank much more than four-tenth of an inch, or stood lower than 29.45 inches. These gales began westward off north along the shore, veered to eastward, and increased in intensity as they went round. This change of direction continued, the wind gradually abated, until it was at south, when it frequently fell almost calm.[66]

John Goldingham, the Dane employed in the Madras Observatory, conducted researches. According to his report of 27 October 1797, the moon having just passed the first quarter, and being at its greatest distance from the earth, there was a violent gale of wind at Madras, somewhat resembling the storms of recent years. It began from northwards in the night between the 26th and 27th, veered to the north-east. In the morning it blew with uncommon force for three hours. About noontime, it suddenly shifted to the south, and it was almost as violent as before. Many old trees were uprooted. The barometer in Madras at about noon on the 25th was at 30.005 inches. At 2 p.m. on the 27th it had sunk to 29.465 inches. Though this was not so violent when compared to the late storms, it was no ordinary gale of wind.

On 10 December 1807, the moon half-way between the first quarter and full and nearly at the greatest distance from the earth, there was a gale at Madras. It began in the evening from the north, and was accompanied by thunder, lightning and rain. It veered to the southward of east, and blew with violence. It slackened gradually in the afternoon, and at 1 p.m. the sky was clear. The barometer fell about 0.40 of an inch. The rainfall was found to be at 3.50 inches.

On the night between 29 and 30 March 1820, a strong gale of wind developed; the moon at the full, but nearly at its greatest distance from the earth. This gale commenced from the north-east and blew with immense force at times. Contrary to the course of the monsoon gales, it veered to the north, north-west, and south-west, still violent. But at the latter quarter it gradually slackened, and broke up at about 9 a.m. The barometer fell at 0.40 of an inch. It was a little above 29.5 inches when at the greatest depression took place. The rainfall was found to be 6.50 inches. The ships left the roadstead in the evening. Some of the smaller crafts were driven onshore. The others went down at their anchors. Several ships and smaller vessels were lost along the coast during the gale. It had been more violent towards the north than at Madras. The storms that occurred at Madras resembled whirlwinds, blowing all round the compass (from particular points with incredible fury). They were confined to a space comparatively small in diameter. Goldingham gave some of the leading features.

He reported that on 2 May 1811, a violent storm occurred at Madras. The moon had passed the first quarter on 30 April, and was full on 8 May. It was also at the greatest distance from the earth. This storm raged with great fury and did considerable damage. The barometer was so much out of order that nothing could be correctly stated regarding the actual quantity of depression. The rainfall was about 5.50 inches. It appeared from the notices published at the time that early on 1 May the surf was unusually high, while thick clouds continued to gather during the day from the north-east. The wind blew very hard by daylight accompanied by heavy rain. At about noon, it increased, and towards midnight it reached its greatest height, when it blew with incredible fury. The storm raged with such great fury that it destroyed every vessel on the roadstead, with the exception of three: a small Spanish ship, an American, and a French cartel ship. These stood out to sea, but the former was driven onshore near Kovalam. 90 country vessels went down at their anchors. All the rest were driven onshore, along with the *Dover*, a frigate, and *Chichester*, a store-ship. The whole beach of Madras was covered with wrecks and dead bodies for 2 miles north and south of the town. The papers stated that the storm was not felt at regions that were at a distance of about 40 miles from Madras.[67]

On 24 October 1818, a second violent storm battered Madras. The moon had passed the last quarter about two days, and it was nearly at its greatest distance from the earth. The wind which was a strong northerly gale early in the morning before 10 a.m. had increased to a storm. A pause of half-an-hour occurred about this time. After which it blew a complete ouragan from the south, with a fury never before experienced at Madras. Some of the oldest trees that resisted during the former storms were uprooted. The largest branches of the trees were torn off by the force of the wind. In some trees of a tough description of wood, such branches were seen hanging down and twisted, having been whirled round and round by the fury of the storm. Several native habitations were levelled. Many of the larger buildings damaged, and some lives were lost. Several ships were at anchor in the roadstead and all those under-weight were driven onshore. One of the ships got foundered. Another was driven onshore to the northward, and a third rendered unseaworthy, while the others generally sustained great damage. The rainfall was about 5 inches. The barometer had fallen between 8 p.m. of the 23rd and the daylight of the 24th, nearly three-tenth of an inch, standing at about 29.5 inches. During the lull at 10 a.m. it was at 28.780 inches, the most extraordinary and terrific depression, such as had never before heard of at Madras. Towards noon, it began to rise and at sunset it was at 29.65 inches.[68]

On 9 May 1820, another storm occurred at Madras. The moon was between the third quarter and new, but at its nearest approach to the earth. The storm began in the evening of 8 May in a gale from the north-west. It increased and blew very strong before morning, accompanied by torrents of rain. Violent gusts continued all day of the 9th, when the wind began to shift to the west and south-west, blowing with greater violence, than before. The rain fell in torrents. Before noon on the 10th, the violence of the storm had subsided. This storm was of far longer duration than either of those that preceded it. The torrents of rain veered to different points of the compass. The damage on shore was great and distressing. A great many lives were lost in the vicinity of Madras. Most of the ships put to sea early and small vessels suffered great destruction.

The *yeris* and *kulams* burst, and the *aarus* overflowed in all directions, destroying property. This storm had a wider range than the former. The barometer on the 8th in the forenoon was at 29.750 inches. It had fallen at sunrise on the 9th to 29.400 inches. By noon, on the same day, it fell to 29.135 inches. At 3 p.m. it fell further down at 28,816 inches and at 5 p.m. it was at 28.670 inches, found lower even than during the former storm. By sunrise, on the 10th it had risen to 29.633 inches. Between the 8th at night, and the 10th at sunrise, 16 inches of rain had fallen.

Goldingham stated that the views of scholars who opined that the moon as having great influence in gales and storms need revision. He said that in his research he noticed as having the least influence. In one gale only the moon was at full, having been at the greatest distance from the earth. In four of the above instances, the moon, upon a mean, passed the meridian at 30 degrees south of the zenith of Madras. In three instances she passed at 6 degrees south and the remaining two within about a degree to the northward. He said that if one were inclined to draw any conclusion from the circumstances under which these gales and storms occurred, it might be, better to ascribe to the moon a protecting power against such visitations, instead of aiding to produce them; having been generally far removed from the earth at the time, and not in that part of its course where, acting in combination with the sun, it might have had the greatest influence.[69]

Goldlingham carefully examined the weather of Madras and he differentiated the days during 1820-1 into four types, namely, cloudy, hazy, clear and thin haze. For the rains, he mentioned the actual days of raining. The English Company records mentioned that ouragans and storms had hit Madras on 30 October 1836, 22 May 1843, 25 November 1846 and in March 1854 and in May 1858 and caused damage to ships and loss of lives.[70]

THE STUDY OF STORMS BY J.J. FRANKLIN IN MADRAS FROM 20 OCTOBER 1846 TO 25 NOVEMBER 1846

The gale which was experienced at Madras on 20 October 1846 was not of that violent nature. Many people imagined that it was not the class of rotatory storms because the wind during its continuance did not veer round more than a point or two. Some hours before it began at Madras, the wind was blowing briskly from west-south-west to west-north-west with the usual rainfall and a sky was of a leaden colour. In the afternoon it veered back to the westward, and gradually drew to west-south-west. The sky became more obscure, and the barometer indicating approaching bad weather.[71]

It then increased gradually till, about 9 p.m. when it was blowing a fresh gale from west-south-west. The rain at the same time began falling in sheets of water. By midnight, it had developed into a moderate ouragan, at which it remained till about 4 a.m. On 21st, the barometer fell to 29.492 inches, and began to rise slowly. At midnight the wind was from the south-west. The indicating pencil of the rain gauge got damaged and the direction of the wind was lost. At 7 a.m. the wind was blowing from south-west and it was observed as extraordinary. The wind remained tolerably steady during the above time when its greatest force was exerted.

The amount of pressure on the square foot during this time was 174 lbs that denominated a heavy gale. The force was greater at sea than on shore. This led to the conclusion that the body of the ouragan was not passing from east to west, but was formed in such a position as to throw both Madras and the shipping in its south-eastern quadrant. At 4 a.m. the wind together with the rain, began to subside. It drew towards south, in which quarter it remained strong and variable till between 7 and 8 a.m. on 22nd. It backed round to the north-west, and the barometer gradually rose till the 25th when it attained its previous height of 30 inches. The ships, *Ann Armstrong*, *Edward Bilton*, *Lady McNaughten* and *Eleanor Lancaster* with passengers on board were put on sea between 30 and 40 miles from land. The barometer at that time ranged 29 inches during the height of the gale.[72] It attributed to the instability of the vessels. In the morning of 21st, it blew at all places from east to north-east at which time, it was still blowing hard at Madras from south-west to south. It then travelled slowly to the northward, and made a curve towards seawards. By about noon of the 22nd, its impact was no longer felt at Madras, the nucleus resumed its course over the land.

The ouragan in Madras was of rotator nature. The violence of the wind was exerted to great extent. The most remarkable feature in it was the extraordinary rainfall that took place during its duration. From the sunset of 20 November 1846 to the following day's sunrise, the pluviometer showed an amount of 17.5 inches rainfall. It was unprecedented in the meteorological annals of Madras. The whole quantity that fell from the sunrise of the 20th to the following day's sunset was 24.33 inches. The ouragan of 25 November was unusual, preceded in Madras by a murky atmosphere with the wind blowing in squalls from the north-west to north with heavy rain. After noon it veered round to north-north-east and north-east, the wind rapidly increased.[73] The storm had formed about 300 miles east by north of Madras. Had it come further from eastward, the ship *Cressey* would have sunk. Between 24 and 25 November 1846, the vessel sailed to the south and the captain was confused. The ship finally struck about midway between Madras and Sadurangapattinam.[74]

At noon of 25 November 1846, J.J. Franklin fixed the nucleus about 160 miles eastward from Madras. The ships that left the roads on 24th began to feel the ouragan from the northward, from which quarter it rapidly veered to west and south-west as the centre approached and it passed to the north. The ship *Agincourt* veered to south-south-west and ultimately to south-east by east as the wind passed through the south-eastern quadrant of the vortex.

The two English ships *Macedon* and *Seringapatam* were both at sea. They experienced a gale between south and south-west, clearly indicating that they had been sufficiently far from the nucleus. Therefore, the captains of the ships did not feel the sudden shifts of wind that always had been there. On this occasion, they found it to be so dangerous to vessels. The captain of the ship *Macedon* subsequently sailed to the north-west and crossed the track of the ouragan, before it had passed and went over the ground. The sailing vessel *Agincourt was* furthest to the eastward, had set sail on the night of 23 November 1846, it therefore, faced the greatest fury of the storm about midday of the 25th. The barometer showed 28.57 inches at that time. The ships *Anna Robertson* and *Eleanor Lancaster* left the roads early on the afternoon of the 24th and experienced its greatest force between 4 and 5 p.m. The ship *Zarah* experienced the height of the storm about the same time as the other two above-mentioned vessels. Owing to a sudden shift of wind it sank and shifting cargo was in a heavy lurch. The wind in Madras then blew from north to north-east.

The barometer indicated foul weather on the morning of 25 November 1846.The remaining ships *Athenian*, *Edward Bilton*, *Augusta*, and *Jane Catherine* left between 11 a.m. and 2 p.m. on that day. They sailed to the east and encountered a lull between 7 and 8 p.m. Previous to the lull the wind blew from north-east to north-north-east. The sky remained as dense as ever but the barometers showed no indication of a rise. It remained calm for about half-an-hour when the blast returned in all its fury from west to south-west, throwing all the three vessels. The *Edward Bilton* and *Augusta* were dismasted—the latter waterlogged and was abandoned—but the *Jane Catherine* providentially escaped without any serious damage.

The centre of the storm passed to the north and their positions had been found between 30 and 40 miles off shore. This was between 7 and 8 p.m. At about 9.45 p.m., the greatest depression of the barometer took place at Madras with 29.03 inches. The centre had then reached its nearest proximity to Madras. The body of the storm passed to the westward, and then to the southward. The gale was felt at Sadurangapattinam where the trees were all blown down.[75]

J.J. Franklin said that no register of the force of the wind could be obtained after 8 p.m. of 25 November 1846 because the connecting link between the registering pencil and the plate got damaged. There was no possibility of replacing them. The force of the wind in Madras was, however, computed by a talented engineer officer. It was found to be 52 lbs. to the square foot. The pillars of the Elphinstone Bridge in Madras at that time were completely blown away.[76] J.J. Franklin had analysed the storm of 1846 and he had given a wide backdrop for the understanding of past climate.

Henry F. Blanford who studied the storms of the sea wrote also on the daily average movement of the wind. He calculated scientifically and noted from the anemometer in Madras the average

TABLE 4.2: REGISTER OF THE SYMPIESOMETER, BAROMETER, AND FORCE OF WIND AT MADRAS, 25 NOVEMBER 1846-26 NOVEMBER 1846 TAKEN BY J.J. FRANKLIN

Day and Time	Sympiesometer	Barometer	Wind
25 November 8 a.m.	29.26	29.70	5.5 lbs
25 November 1 p.m.	29.10	29.57	7.5 lbs
25 November 2 p.m.	29.06	29.54	8.5 lbs
25 November 3 p.m.	29.06	29.52	8.5 lbs
25 November 4 p.m.	29.04	29.50	8.0 lbs
25 November 5 p.m.	29.01	29.49	9.5 lbs
25 November 6 p.m.	28.98	29.46	14 lbs
25 November 7 p.m.	28.91	29.40	19 lbs
25 November 8 p.m.	28.85	29.31	27 lbs
25 November 9 p.m.	28.62	29.12	–
25 November 9.30 p.m.	28.58	29.03	–
25 November 10 p.m.	28.58	20.03	–
25 November 10.30 p.m.	28.70	29.16	–
26 November 5.30 a.m.	29.26	29.63	–
26 November 8 a.m.	29.36	29.73	–

TABLE 4.3: DAILY MOVEMENT OF THE WIND IN MADRAS AND NAGAPATTINAM IN 1885

Month	Madras	Nagapattinam
January	150	132
February	122	83
March	152	82
April	191	114
May	224	164
June	218	180
July	202	175
August	179	142
September	159	126
October	123	88
November	166	126
December	184	163

Source: Henry F. Blanford, *A Practical Guide to the Climates and Weathers of India, Ceylon and Burma, and the Storms of Indian Seas,* London: Macmillan, 1889, p. 31.

daily registered wind that was at 50 miles. He mentioned that at Nagapattinam the wind was only at 27 miles. Thus, progress was made in the study of winds, gales, storms and cyclones through instruments like barometer, sympiesometer and anemometer brought from Europe to Madras.

CONCLUSION

It is concluded that winds in the pre-monsoon period moved generally slow. The storms of the monsoon period normally attained great violence and caused terrific winds and so heavy rainfall occurred on the Tamil coast. They caused damages to boats, ships, houses and buildings, besides pulling down magazines and hospitals. John Goldingham in Madras conducted the scientific study of gales and storms between AD 1797 and 1820. J.J. Franklin in Madras studied the storms of 1846. It was found by them that storms had travelled towards the east coast of peninsular India and that the frequency of occurrence was greatest in November. Further, storms crossed the coast to the north of Madras and also to the south of Madras. Nearly 50 per cent of the storms that originated in the Bay of Bengal travelled and crossed near Madras. Each storm and each cyclonic event in Madras and Pondicherry had illustrated a specific development in the ways in which the English and the French colonial administrators responded to crises in the colony. Ananda Ranga Pillai of Pondicherry, in his private diary in Tamil, often furnished details of storms which are highly descriptive. As a writer he was clearly moved by the plight of men rendered powerless by the forces of nature. The calendar system followed by the Portuguese and French (Catholics) differed from that of the Dutch, Dane and English (Protestants) on the Tamil coast. The Dutch, the English and the Danes used the Julian calendar, which was 11 days behind (not adopted by the English until AD 1752) the Gregorian calendar of the Catholics. Hence, a possible discrepancy must be avoided in the course of research with reference to the exact dates of storms and cyclones that occurred.

NOTES

1. P.S. Sundaram, trans. & ed., *The Kural*, New Delhi: Munshiram, 1990, chapter II, no.17, p. 12.
2. N.M. Venkatasamy Nattar and R. Venkatachalam Pillai, eds., *Aganaanuru*, 3rd edn., Madras: South India Saiva Siddhantha Pathippu Kazhagam, 1957, p. 255.
3. K.A. Nilakanta Sastri, 'Takuapa and its Tamil Inscription', in *Journal of the Malaysian Branch of the Royal Asiatic Society*, vol. XXII, 1949, pp. 25-30.
4. Armando Cortesao, ed., *The Suma Oriental of Tome Pires and the Book of Rodrigues*, 2 vols., New Delhi: Asian Educational Services, 1990, vol. I, p. 66.
5. British Library [hereafter BL], London, *Original Correspondence*, no. 1748.
6. William Foster, *The English Factories in India: A Calendar of Documents in the India Office, British Museum and Public Record Office* [hereafter *EFI*], XIII vols., Oxford: Clarendon Press, 1906-17 vol. IX, p.178.
7. Ibid., vol. XI, 12 May 1662.
8. BL, *Original Correspondence*, no. 3247, 23 January 1668-9.
9. Love, *Vestiges of Old Madras: Traced from the East India Company's Records Preserved at Fort St. George and the India Office and from Other Sources*, 3 vols., London: 1913, reprint, Delhi: Asian Educational Services,1996, vol. I, pp. 477, 479-80.
10. Ibid., p. 480.
11. Ibid., pp. 481-2.
12. Ibid., p. 330.
13. Ibid., vol. II, p. 177.
14. Ibid., pp. 193, 203, 206.
15. Ibid., p. 230.
16. Marc Vigié, *Dupleix*, Paris: Fayard, 1993, p. 244.
17. R. Walsh, R. Glaser and S. Militzer, 'The Climate of Madras during the Eighteenth Century', in *International Journal of Climatology*, vol. 19, issue 9, July 1999, pp. 1025-47, see p. 1039; Gnanou Diagou, ed., *Prathiyegamana Ananda Ranga Pillai Avargalin Sostha Ligitha Thinappadi Seythikurippu*, 8 vols., Pondicherry: Government of Pondicherry, 1948-54; Alalasundaram, ed., *Ananda Ranga Pillai Naatkurippu*, vols. 9-12, Pondicherry: The Author, 2005; Frederick Price, H. Dodwell and V. Rangachari, trans & ed., *The Private Diary of Ananda Ranga Pillai* (hereafter *ARP Diary*), 12 vols., reprint, Delhi: Asian Educational Services, 1980. The Tamil version of *ARP Diary* contains the entry dated 14 October

1746. The English translation provides only a summary. *ARP Diary*, vol. II, pp. 301-2.
18. A.T. Mackenzie, *Official Papers Concerning the Construction of the Madras Harbour*, Madras: Government Press, 1902, pp. 4-5.
19. R. Walsh, R. Glaser and S. Militzer, *The Climate of Madras*, p. 1039.
20. The Tamil version of *ARP Diary* contains the entry dated 1 November 1752, which the English translation has omitted.
21. R. Walsh, R. Glaser and S. Militzer, *The Climate of Madras*, p. 1039.
22. Henry Davidson Love, *Vestiges*, vol. II, p. 567.
23. Ibid., p. 620.
24. A.T. Mackenzie, *Official Papers*, pp. 4-5.
25. R. Walsh, R. Glaser and S. Militzer, *The Climate of Madras*, p. 1039.
26. Henry Davidson Love, *Vestiges*, vol. III, p. 219.
27. Ibid., p. 201.
28. Alfred Martineau, 'Les cyclones â la cote Coromandel de 1681 â 1916', in *Revue Historique de l'Inde Française*, vol. II, 1917, pp. 229-324, see p. 239.
29. R. Walsh, R. Glaser and S. Militzer, *The Climate of Madras*, p. 1040.
30. Georg Christian Knapp, et.al., *Neuere Geschichte der evangelischen Missions-Anstalten zu Bekehrung der Heiden in Ostindien aus den eigenhändigen Aufsätzen und Briefen der Missionarien erausgegeben*, [hereafter NHB-Neue Hallesche Berichte], Teil 1-8, (Stück 1-95), Halle: Waisenhaus, 1770-8/ 1795-1848, NHB 50. St., 1238-1309.
31. A.T. Mackenzie, *Official Papers*, pp. 4-5; Henry Davidson Love, *Vestiges*, vol. III, p. 450.
32. Archiv der Franckeschen Stiftungen [hereafter AFSt], Halle: AFSt/M2 B2:14a, *Meteorologische Observationen vom Jahr 1789*; AFSt/M2 B2:17, *Meteorologische Observationen vom Jahr 1791*.
33. AFSt/M2 B2:14b, Meteorologische Observationen vom Jahr 1790.
34. John Goldingham, 'Results of Meteorological Inquiries, Made at Madras', in *Transactions of the Royal Asiatic Society of Great Britain and Ireland*, vol. 3, no. 3, 1834, Appendix no. III, p. xxx.
35. R. Walsh, R. Glaser and S. Militzer, *The Climate of Madras*, pp. 1039-40.
36. NHB 18. St., 668-70; NHB, 48. St., 1089-99.
37. *Revue Coloniale*, Paris: Imprimerie et Librarie Adminsitative de Paul Dupont, tome XVIII, 1857, p. 53.
38. R. Walsh, R. Glaser and S. Militzer, *The Climate of Madras*, pp. 1039-40.
39. A.T. Mackenzie, *Official Papers*, pp. 4-5.
40. Alfred Martineau, *Les cyclones*, p. 236.

41. Gnanou Diagou, *Prathiyegamana Ananda Ranga Pillai*, see entry dated 23 October 1745. The English translation has omitted this entry.
42. Ibid., entry dated 30 October 1752. The English translation has omitted this entry. *ARP Diary*, vol. IX, pp. 101-2. The entry of 8 November 1754 mentions the rain of 6 November 1754. *ARP Diary*, vol. IX, p. 90.
43. Alfred Martineau, *Les cyclones*, p. 231.
44. Yvonne Robert Gaeble, *Une Parisienne aux Indes au a XVIIIe Siecles*, Pondicherry: L'Imprimerie du Gouvernement, 1958, pp. 68-9.
45. Ibid., p. 83.
46. *ARP Diary*, vol. I, pp. 289-91.
47. Gnanou Diagou, *Prathiyegamana Ananda Ranga Pillai*, p. 219.
48. *ARP Diary*, vol. I, pp. 289-91.
49. Ibid., p. 220.
50. *ARP Diary*, p. 293.
51. Ibid.
52. Ibid., vol. I, pp. 292-3.
53. Ibid., p. 292.
54. Alfred Martineau, *Les cyclones*, p. 232.
55. Gnanou Diagou, *Prathiyegamana Ananda Ranga Pillai*, entry dated 23 October 1745.
56. G.B. Malleson, *History of the French in India: From the Founding of Pondicherry in 1674 to the Capture of that Place in 1761*, London: Longmans, Green & Co, 1909, p. 573.
57. Alfred Martineau, *Les cyclones*, pp. 236-8.
58. G.B. Malleson, *History of the French in India*, p. 573.
59. Ibid.
60. Alfred Martineau, *Les cyclones*, pp. 239-41.
61. Ibid., pp. 240-1.
62. A.P. Dubois and Xavier Raymond, *Inde*, Paris: Poulet, 1845, p. 442.
63. Edward Thornton, *The History of the British Empire in India*, London: Longmans, 1859, p. 78.
64. Alfred Martineau, *Les cyclones*, p. 240.
65. Ibid., pp. 240-1.
66. John Goldingham, 'Results of Meteorological Inquiries Made at Madras', p. xxxii.
67. Ibid., p. xxxiii.
68. Ibid, pp. xxxiii-xxxiv.
69. Ibid., p. xxxiv.
70. A.T. Mackenzie, *Official Papers*, pp. 4-5.

71. J.J. Franklin, 'Notice of the Storms by J.J. Franklin', in *Madras Journal of Literature and Science*, vol. XIV, 1847, pp. 146-51, see pp. 146-7.
72. Ibid., p. 147.
73. Ibid., p. 148.
74. Ibid., p. 149.
75. Ibid., p. 150.
76. Ibid., p. 151.

CHAPTER 5

Hazards of Sea, Land and Water: Floods, Tsunamis and Earthquakes, Ninth-Nineteenth Centuries

The south-west and the north-east monsoons had brought rains to the Tamil coast and the availability of water in abundant supply was only from nature. Medieval Tamil evidences mention heavy rainfall and the unprecedented rains that caused floods. The inundation of paddy fields through flooding during this period caused distress to both peasants and the public in general.

THE HEAVY RAINS, UNPRECEDENTED FLOODS AND SOIL EROSION IN MEDIEVAL TAMIL COUNTRY

Floods in Kaveri river laid waste a large area of land as found mentioned in inscriptions.[1] The records mention floods that occurred in the Srirangam area.[2] Plots of land in the places therefore, had become uncultivable on account of the sand and silt that was caused by the overflowing of the Kaveri. It caused considerable damage to the paddy fields and other crops.[3] Damage to the lands in the villages of Allur and Isanamangalam, located on the southern bank of River Kaveri, had been reported during the rule of Rajendra Chola I (AD 907-1044). Efforts were taken by the villagers to reclaim the lands by sand cast.[4] The floods in Kollidam river also often resulted in the dislocation of land boundary stones. In one instance, we find that the boundaries of the lands held by two temples at Srirangam and Thiruvanaikaval remained unsettled for a long time.[5]

Floods in Palar river in North Arcot and Chingleput area have been mentioned. The cultivating lands in the village of Thirumalpuram in AD 1004 and Puthupaadi in 1071 had been affected.[6] A massive flood at Thiruvottur in North Arcot area affected peasants in 1125.[7]

TABLE 5.1: THE PERIODICITY AND FREQUENCY OF FLOODS IN THE LOCALITIES OF MEDIEVAL TAMIL COUNTRY, AD 937-1402

Year AD	Locality	Region	Year amongst the 60 Years of Tamil Calendar
937	Allur	Tiruchirapalli	Yevilambi 31st year
937	Isanamangalam	Tiruchirapalli	Yevilambi, 31st year
1004	Tirumalpuram	North Arcot	Khrodhi, 38th year
1071	Puthupaadi	North Arcot	Virodhikrithu, 45th year
1125	Tiruvottur	North Arcot	Visuvaavaha, 39th year
1126	Tiruvottur	North Arcot	Parabhava, 40th year
1157	Vijaynarayanam	Nanguneri	Isvara, 11th year
1175	Tiruchi	Tiruchi	Manmatha, 29th year
1190	Somangalam	Chingleput	Sadharana, 44th year
1191	Somangalam	Chingelput	Virodhikirithu, 45th year
1402	Valuvur	Thanjavur	Chitrabhanu, 16th year

Source: S. Jeyaseela Stephen, 'Rainfall and Irrigation in Medieval South India and the Socio-Economic Impact', in *Proceedings of the Indian History Congress*, 66th Session, Delhi, 2006, pp. 296-308; S. Jeyaseela Stephen, 'Water Resources and Management in the Rural Setting of Medieval South India', Peter Borschberg and Martin Krieger, eds., *Water and State in Europe and Asia*, Delhi: Manohar, 2008, pp. 105-20.

It stated that this devastating flood had destroyed both the village and crops. Hence, villagers were unable to pay the annual revenue and taxes. As a result, they were forced to sell some wetlands to settle tax arrears of 1125. Again, villagers had to sell their dry lands in 1126 for the same purpose.[8] Heavy rains led to flooding in Chingleput region in 1190 and 1191 and the breach of the *kulam karai* at Somangalam took place.[9] At Thiruvanaikaval, heavy floods in Kollidam river caused extensive breaches to the *karais*. Silting of wetlands in the villages of Kannanur, Narasingamangalam, Sengavur and Ottanur occurred in AD 1258. These lands were reclaimed after a long time.[10]

FLOODS AND THE ROLE OF THE MEDIEVAL STATE

Rulers of Tamil country realized the impact of flood-induced damages. Hence, they took necessary steps and strengthened the earthen embankments to prevent and control floods. *Mathugus* were also built to act as spillways to relieve excess pressure during flood and the top of the banks were planted with grass to mitigate erosion. An order was issued by Kulothunga Chola (AD 1070-1120) to make paving stones by digging earth and create fresh banks since the *kulam* was spoiled during floods with broken banks. The King donated some garden land and wetlands in the hamlet of Nettaipakkam for meeting the expenses. A royal official was appointed to look after the work. The village assembly of Thiru-kanchi had resolved to assign the income of two villages to repair the *kulam* damaged by floods in the near future and also use the incomes for any repair.[11]

Flooding of the Kaveri river had damaged the cultivating lands through Tandurai Jayankonda Vaykaal. Therefore, steps were undertaken for the reclamation of lands during the rule of Vikrama Chola (AD 1118-35).[12] At Tiruvottur, a flood in AD 1125 destroyed crops, leading to the sale of some lands by the villagers for raising money to pay the taxes of the year. An exemption of tax was granted at that time by the ruler to villagers since floods had caused heavy damage.[13]

We also find the remission of taxes granted by the Pandya rulers during floods. Maravarman Srivallabha (AD 1145-62) in 1157 came forward to extend help when there was unexpectedly heavy rain in the month of *Masi* (February-March) which caused floods. It has been stated that Naomayan yeri, the big *kulam* of Vijayanarayana Chaturvedi Mangalam, had breached the west of the western *mathugu*. There were 32 breaches in the supply *vaykkal* to the *kulam*. The water flows of summer rains were too heavy that year. The *kulam* had not been repaired to receive adequate water supply. Nambi Sankaran Vasudevan of Turuthimangalam the official had neglected his duty to undertake repairs well in advance. The *kulam* was largely fed by a few streams taking off from the Nanguneri uplands. A breach was also beyond the capacity of the *mahasabha* collective action. Through the help of the ruler, the breaches were closed and the *kulam* was strengthened. The ruler also participated in the proceedings of the *sabha* and he granted the remission of taxes in the flood-affected area.[14] Again in 1175, Jatavarman Kulasekhara (AD 1162-77), the Pandya ruler, also granted remission of taxes for five years since the area was heavily flooded.[15]

The occurrence of heavy rainfall at Valuvur in Thanjavur area during November and December 1402 has been mentioned. The Kaveri overflowed its banks and washed away the demarcation boundary stones in the paddy fields. Lands submerged on account of floods in Kaveri were laid waste; the *aaru* had silted the irrigation *vaykkals* and as a consequence the tenants abandoned the cultivation fields. The tenants were granted concession in payment of taxes at that time.[16]

EUROPEANS ON THE TAMIL COAST: REPORTING FLOODS AT NAGAPATTINAM IN AD 1681 AND AT MADRAS IN AD 1682

Nagapattinam, the Dutch settlement, saw massive inundation in November 1681. Strong winds struck 70 miles across the coast, affecting the coastal people. As many as 2,000 fishermen took shelter in a big temple that stood on high ground but the entire structure—unable to withstand the fury of the seas—collapsed upon them, killing them all. The total deaths in Nagapattinam area alone were more than 14,000 people.[17]

In the English settlement of Madras, floods were reported on Tuesday, 11 July 1682. The sea had encroached upon the town ten days ago. The English hoped that it would retreat as usual. Therefore, they did not make any serious effort. Soon they realized that the walls would collapse being under water for long. They thought Madras would be swallowed up. They tried to protect the very foundation of the town. The English Company officials feared that the great bulwark next to the seaside would also yield to its fury. It was ordered immediately to call all workmen, such as coolies, carpenters, smiths and peons for work. Sufficient materials were collected and given and the work was undertaken day and night.[18]

REPORTS OF FLOOD IN TRANQUEBAR IN AD 1710, 1745 AND IN PONDICHERRY IN AD 1745

European travellers and missionaries in Tamil country observed the annual meteorological events and natural calamities such as occurrence of floods and other extreme weather events. The Lutheran missionaries in Tranquebar reported many events of floods and said that those of 1710 and 1745 were directly linked to tropical cyclones and produced both coastal inundations from sea and river flooding.

Jean François Charpentier de Cossigny, a military engineer, was employed in Pondicherry. He planned to fortify the town during 1734-41, by reinforcing with masonry the defensive ditch at the Villiyanoor Gate in the south-west quarter of the town. This he aimed to undertake in order to prevent future flooding, but the officials in Paris refused to fund the project.[19] Cossigny identified Villiyanoor Gate as the weak point in the colony, left open to flood defenses. When floods occurred in Pondicherry in 1745, Cossigny was blamed for the fury of the flood that ravaged the native quarters, though funding was not earlier provided by the directors of the French East India Company in Paris. It is mentioned that more than 2,000 houses in the French colony were swept away during the flood. 40 people perished in the water that inundated the black town.[20] This event was reported by an officer in Pondicherry to the directors of the French East India Company in Paris. This is the first time we notice an individual—Cossigny—being blamed for the flood havoc.

When we go through the diary entry of Ananda Ranga Pillai in Pondicherry, we find that he had mentioned that the banks of the Uppar river had overflowed and the neighbourhood area houses were swept away. He said that the *mathugu* had given way and having burst through the *mathugu*, the flood ran through the town. He expressed that these *mathugus* were built for the purpose of controlling the flow of water into the town by diverting water into surrounding *vaykkals.*[21]

REPORT OF FLOODS IN CUDDALORE IN AD 1754

The Lutheran missionaries very carefully noted and described the passage of cyclones directly over Cuddalore with northerly winds and followed later by southerly winds on 14 April 1749.[22] The flood damage in Cuddalore in 1754 was caused by the Kaveri river which had its headwaters in the *ghats*, and its downstream in

River Kollidam. The most severe and widespread river-flooding, occurred in November-December 1754, and the destruction and heavy loss of life caused by it had been reported in detail by Johann Kiernander and Conrad Huttemann, two Lutheran missionaries, in their letters and annual report. According to them, the worst damage was caused in part by the bursting of a wall of a *anaikattu* in one of the reservoirs on the floodplain, half-a-mile away from Cuddalore on the afternoon of 15 November 1754. In a letter dated 20 November 1754, the missionaries stated that after continuous rains during day and night for more than 20 days, all the small *yeris* and *kulams* (here denoting ponds) became so full that the *anaikattu* walls broke. This resulted in such an extraordinary flood that the whole area around Fort St. David at Cuddalore was under deep water. The new settlement of Puthu Palayam at Cuddalore was almost completely destroyed. In Vandi Palayam and Tirupapuliyur, several houses were swept away.[23] River floods had also been reported by the Protestant missionaries in Madras in 1752 and in 1754 in the whole of Tamil country, and at Cuddalore in 1761 at Tranquebar in 1772, 1776 and 1794 because of heavy rains during the north-east monsoon.

FLOODS IN TRANQUEBAR IN AD 1771, 1775 AND 1778

It has been mentioned that the river-flood of AD 1778 in Tranquebar was caused by a tropical cyclone. Two instances of river-flooding were reported as occurring during the summer south-west monsoon period, while that of 1771 coincided with exceptionally wet conditions at Tranquebar. The flood in Nandalaru, a branch of the Kaveri in 1775 was a result of heavy rain during July-August in the interior. It is evident from the descriptions of events by the Lutheran missionaries that flooding at the various port settlements of Cuddalore, Tranquebar, Nagapattinam and Porto Novo had been owing to three different causes. The first type included coastal inundations related to tropical cyclone events or other high winds. The second was due to river-flooding linked to local heavy rain or cyclonic events. The third was the exceptional flooding of major *aarus* linked to heavy rain in the region over their headwaters in the *ghats.*

REPORTS OF FLOOD IN THANJAVUR IN AD 1784, 1812, 1826, 1831 AND 1843

Eighteen cyclones had hit the Coromandel Coast between AD 1710 and 1799, an average of two cyclones per decade. The frequencies varied, with the missionaries reporting five in the 1740s and four in the 1780s. They have also mentioned severe drought in Tranquebar during 1726-9 and the Great Famine of Madras of 1769-70. Sir Thomas Herbert, an English official, remarked that at Nagapattinam the climate was very hot and the winds were strong. He said that during some months of the year the heat was so high and unseasonable that people were vexed with fever, fluxes and other diseases.[24]

The north-east monsoon of 1754 resulted in massive river-flooding and damage to house and property in the whole region. In 1778, it was described as the wettest and longest north-east monsoon season in 40 years, and it had led to great flood damage. The south-west monsoon rains were noted as anomalously wet in 1710, 1768, 1771 and anomalously dry in 1734, 1737, 1754, 1761, 1770, 1773 and 1794. The reports of dry years in 1726-9, 1740, 1746, 1774 and 1788 were not specific as to whether the dryness applied to the rainy seasons too. Floods had occurred in Thanjavur in 1784, 1812, 1826, 1831 and 1843.[25]

FLOOD IN PONDICHERRY IN 1830 AND THE FLOOD RELIEF COMMITTEE APPOINTED IN 1831

French and English records neatly mention the storms that affected so numerously and there were nine storms that occurred during the English occupation of Pondicherry between 1793 and 1816.[26] After the restoration of Pondicherry to the French in 1816, the major cyclone to strike the town was in December 1830. De Mélay, the Governor of Pondicherry, thereafter called the Council in Order to discuss what measures should be taken by the government for the immediate relief of the poorest inhabitants, a great many who were without shelter in their ruined huts. The Governor explained how the necessary expenses of flood relief would likely impact the revenue. He said that they should have put in effect the orders of the ministry, especially as regards the current discrepancy between receipts and expenses, which only aggravate the difficulty of their position. In his letter dated 29 January 1831 to the Foreign Minister in Paris, the Governor mentioned the storm and the floods had destroyed a large part of the crops, nearly all fruit trees, and damaged the houses within a radius of six miles around Pondicherry. The botanical garden in the town was completely devastated and the promenade, recently completed at great cost, was littered with debris torn off the sea wall by the double action of the heavy rain and the sea.[27]

The flood relief measures were announced by the French administration through a three-man committee appointed to distribute Rs. 500 in aid for the reconstruction of Indian houses of the meanest stature. A further sum of Rs. 108 was dispersed equally among the 54 inhabitants of Thengathittu, the island of coconut plantations inside the mouth of the Chunnambar river. Nearly 500 damaged trees were felled and removed but, oddly, the cost of their removal was the responsibility of the island's inhabitants. Each of them had been given the money to rebuild their destroyed homes. Affected populations engaged in vocations necessary to trade were assisted on credit. Weavers in Orleans Puram and Desbassysns Puram were advanced nearly 500 *pagodas* to rebuild, 12 *pagodas* per house, over twenty times the amount given to the poorer plantation workers south of the town. The repayment of the amount given to the weavers was guaranteed by the privately owned textile manufacture, Poulain and Company.

Salangai boat owners, those who plied the cargo between the ship and seashore in small *masula* boats were advanced 100 *pagodas* to rebuild their launches as well as given a one-time relief of Rs. 46 each, slightly more than that given to weavers. Betel growers were also given aid, while non-betel farmers' requests for relief were rejected on the grounds that damages totalled less than half the value of their farms. A plantation owner who lost 94 per cent of his coconut and 83 per cent of his total crop was only given a tax remission of 64 per cent. Necessary financial support was expected from Paris and it was decided that further steps would be taken by the local administration.[28]

After having temporarily recovered from the effect of the cyclone and flood of 1830, an order was passed by the Governor to regulate the charitable contributions for the amelioration of the poor. The text of the order cited as its motivation the inevitable fraud likely to result from the contravention of fair distribution of aid by the government. The order specified the rates of duty on grains, salt, alcohol, aaru passage enumerated by species and number, and littoral shellfish and shark fin in order to fund monetary efforts at relief.[29] The order also specifically guaranteed that complaints of fraud would not be summarily dismissed by the court because of defects in the complainants' paperwork, providing that there was sufficient proof to proceed. This provision had been included in an effort to make such complaints easier to file by those most likely to be the subject of such frauds, namely those poorest and the least educated Indians who were unable complete formal procedures in French.[30]

FLOODS IN PONDICHERRY IN 1842, 1867, 1872, 1884 AND THE RELIEF WORKS OF THE FRENCH COLONIAL GOVERNMENT

A storm on 18 November 1842 in Pondicherry resulted in floods. The committee appointed undertook the damage surveys and the distribution of relief. The French government spent nearly 60,000 francs on relief and repair during floods between 1867 and 1873. The Minister of the Navy and the Colonies from Paris wrote a lengthy letter to Pierre Aristide Faron, the French Governor in Pondicherry appreciating his efforts at flood relief. He was happy to see that the Governor had found the necessary means within the regular operating budget of the French colony to repair the roads and buildings damaged by the cyclone. He said that the current state of the budget in Paris could not allow relief aid to be allocated from the public treasury.[31]

Torrential rains in 1884 had flooded Pondicherry and Karaikal, causing never-before-seen levels of damage. Property damage was estimated at over 800,000 francs. A sum of 180,000 francs was allocated for immediate flood relief aid to individuals. An additional 57,000 francs was allocated to parish relief subsidies. A telegram dated 22 December 1884, was sent from Pondicherry with the decision of the Council of Ministers to ask the National Assembly in Paris for a credit of 500,000 francs for the aid of the French population in Pondicherry and Karaikal in the terrible trial of flood. Admiral Peyron in Paris communicated that after receiving the telegram the credit was immediately approved by the assembly. It demonstrates the sentiments of solidarity with which the authorities in Paris felt for the colonies of Pondicherry and Karaikal.[32]

It may be mentioned that private charity in Pondicherry also came from five Indians (including two *chettiars*, two *dubashes* and a municipal councillor) and they made sincere efforts at relieving the flood distress of Pondicherry. The mayor of Ozhugarai also did his best with great zeal and he distributed food among the residents of Pondicherry, spending from his own resources.[33]

THE DESCRIPTION AND DAMAGE OF FLOODS IN CUDDALORE IN 1864, 1884, 1898 UNDER THE BRITISH COLONIAL RULE

There was heavy rainfall of 10 inches in Cuddalore from 17-19 October 1864. The Gadilam river flowing through the town, overflowed and swept away the eight arches of the bridge that connected Manjakuppam and Tirupathiripuliur. The Manjakuppam grounds situated at the centre of Cuddalore town was flooded. Patients in the hospitals were quickly transferred to the veranda of the district court because of the incoming flood water. Two women were washed away by the flood and drowned very close to the gate of the collector's bungalow, where the water had come up to the doorway of the building. In Puthu Palayam nearly all the huts were washed away and hardly a house escaped damage. A large number of people were left homeless. A sum of Rs. 3,700 was immediately spent in relieving them. The areas of Villupuram and Tirukkoyilur in the hinterland of Cuddalore had also been severely affected by flood and the havoc was rather considerable. Roads were broken into pieces. Three *anaikattus* on Gadilam river were full and the Perumal kulam, Wallajah kulam and 60 other smaller *kulams* had breached in the area. The estimated amount of the repairs had been said to be over Rs. 1 lakh.[34]

Cuddalore was affected again by a flood in November 1884 and the total rainfall in the region from 4-7 November was 32 inches. The registered rainfall at Cuddalore on 7 November alone was 11.75 inches; it was one of the worst floods ever known. As a result, the Ponnaiyar river, Gadilam river, Vellar river and Gingee river overflowed. The *mathugu* at Edaiyar river, constructed to keep the Ponnaiyar from overflowing, could not withstand the fury of the flood. The Malattar was also in flood and its riverbank had breached. The Ponnaiyar river swept down the Malattar into the Gadilam, which was already swollen with a far greater quantity of water than it could carry; hence, the two *aarus* overflowed. Railway lines between Cuddalore and Port Novo were damaged at three places. The roads in the eastern part of Cuddalore were washed away.

Another flood followed on 17 December 1884 with heavy rainfall registering 15.40 inches. The New Town of Cuddalore suffered greatly. The Gadilam overflowed to its right bank just above the railway bridge near the town and poured into Tirupathiripuliyur by way of a strip of low-lying ground. Many houses were destroyed, and some people died. The four arches of the road bridge connecting Manjakuppam and Tirupathiripuliyur over the Gadilam river collapsed.[35]

Rains continued during the period of 17-19 December and the registered rainfall on 20th December alone was 25.55 inches. The heavy flow of water in the two *aarus* of Ponnaiyar and Gadilam from the afternoon of the 18th till the evening of the 19th had swept through Manjakuppam and Puthu Palayam. With floods, the *mathugus* of the Edaiyar river collapsed. The flood water from the Ponnaiyar river rushed into the Gadilam. The flood water in the Manjakuppam ground stood at 5 ft deep. The flood water at Tirupathiripuliyur was 3.5 ft deep. 11 people died. Nine arches of the Gadilam river bridge were destroyed. 13 arches of the

Ponnaiyar river bridge broke down and disappeared.[36] Gordon, the South Arcot district collector, reported that 956 irrigation sources, 177 imperial and 776 minor works had breached. 20 bridges (large and small) were wrecked. 13,595 houses were damaged. 13,724 cattle, including sheep and goats had perished and drowned.

An embankment was built at that time from the edge of Mount Capper near the Tiruvanthipuram anaikattu to keep the Gadilam river within confines. This was executed to keep the damage to the minimum at Tirupathiripuliyur. The Edaiyar anaikattu was also repaired and rebuilt to keep the Ponnaiyar river from spilling into the Gadilam. The four ruined bridges between Cuddalore New Town and Pondicherry were reconstructed.

Heavy rains pelted the region in October, November and December 1898. Cuddalore town was affected since the Ponnaiyar and Gadilam rivers began to overflow. Serndhanur, a hamlet of Malaiperumal Agaram in Cuddalore area, was underwater for many days. The place could not be approached because of the flood. The Manimukta river overflowed its banks at Vriddhachalam, and so part of that town was inundated. The Raja vaykkaal and Khan Sahib's Canal in Chidambaram town had breached. The Vellar river also overflowed. However, there was no abnormal flow of water in the Kollidam river. It has been mentioned that 87 major irrigation works and 232 minor irrigation works in the area were damaged. The district collector announced tax remissions for peasants because of the resultant damage.[37]

THE TSUNAMIS IN TAMIL COUNTRY IN AD 1788 AND 1881

There are some evidences of tsunami inundation on the Tamil coast. In AD 1788, British explorer William Chambers attributed part of the submergence of ports on the Tamil coast to the overflowing sea caused by an earthquake and the event is described as an earthquake that may have caused the sea to overflow its boundaries.[38] The actual cause of submergence of temples at Mahabalipuram had been attributed to processes related to coastal erosion.[39]

On the morning of 31 December 1881, a submarine earthquake beneath the Andaman Islands generated a tsunami with a maximum crest height of 0.8 m that was recorded by eight tide gauges surrounding the Bay of Bengal. At Madras, a clock in the office of the master attendant, electronically controlled from the astronomical observatory, was stopped at 07.05.45 local time or 07.55.36 Port Blair mean time. Richard D. Oldham (1858-1936), the British official wrote an account of this earthquake and he said that the tide gauges recorded a maximum amplitude of 0.8 m (it was of Mw = 7.9).[40]

It may be mentioned that this earthquake on 31 December 1881, occurred eight years before the construction of the world's first tele-seismic-recording seismometer in the world. Therefore, little is known about its rupture parameters or location. However, it is reported by scientists now that Richard D. Oldham mistakenly located the event deep in the northern Bay of Bengal, based largely on timing data from clocks in Calcutta (now known as Kolkata) and Madras. The island of Car Nicobar was believed to have been raised and tilted during the 1881 event. Deformation models that do not include this uplift result in an inappropriate estimate of the observed tsunami run-up on the island.[41] The earthquake was believed to have occurred on the interface between the Indian and Andaman Plates. The inferred mechanism of westward slip of the hanging wall slip was consistent with the slip partitioning between the dipping subduction zone. The strike-slip east was also consistent with that of Car Nicobar. This earthquake caused minor damage in the Andaman Islands and generated a tsunami that was observed throughout the Bay of Bengal but not along the coast of Myanmar. It must be noted that the tsunami did no damage around the Bay of Bengal. However, it was felt at Ootacamund.

THE EARTHQUAKES IN TAMIL COUNTRY

Nicolau Manucci, an Italian traveller, referred to an earthquake that occurred in Gujarat on 4 January 1699. He mentioned that many Hindu traders from Madras, who were staying there, suffered the misfortune of being buried under the ruins.[42] An earthquake in Madras in AD 1725 and another one was felt at Madras at 9 p.m. on 30 October 1737.[43]

Reportage of earthquakes from Tamil country prior to the eighteenth century had been sparse. Later, English East India Company officials and intellectuals reported earthquakes by colonial observers which became more verbose and eloquent. However, collecting intensity data and initiating studies of specific earthquakes had not begun and developed. Hence, we are unable to know how earthquakes had caused tremors, human sufferings and miseries to the people.

TABLE 5.2: EARTHQUAKES IN TAMIL COUNTRY, 1807-1900

Date	Locality
9 December 1807	Poonamalee/Avadi
10 December 1807	Chennai area
16 September 1816	Chennai
29 January 1822	Vandavasi
2 March 1823	Sriperumbudur
3 January 1859	Kadaladi/Polur
17 December 1859	Salem/Tirupattur
20 January 1860	Shervaroy Hills
2 February 1860	Tirupati
24 June 1865	Coimbatore
3 July 1867	Villupuram/Valavanur
17 June 1879	Hosur
28 February 1882	Ootacamund
12 August 1889	Chennai
8 February 1900	Coimbatore

Source: R. Baird-Smith, 'Memoir on Indian earthquakes, Part I', in *Journal of the Asiatic Society of Bengal*, vol. 11, 1843, pp. 1046-54; Part II, vol. 12 (2), New Series, 136, pp. 1029-56; Part III, vol. 156, 1884, pp. 964-83; T. Oldham, 'A Catalogue of Indian earthquakes', in *Memoir of Geological Survey of India*, vol. 19, 1883, pp. 163-215; A. Bapat, R. Kulkarni, S. Guha, *Catalogue of Earthquakes in India and Neighbourhood from Historical Period up to 1979*, Roorkee: Indian Society of Earthquake Technology, 1983; N. N. Srivastava and K. Ramachandram, 'A New Catalogue of Earthquakes for Peninsular India during 1839-1900', in *Mausam*, 36 (30), 1985, pp. 351-8.

CONCLUSION

It may be said that the disastrous monsoon and heavy inflow of water in *aarus* caused huge loss to life and property. It had posed a major problem to potential agricultural land lost due to soil erosion. The peasants expecting a good *samba* harvest suffered because of the heavy rains and floods during the months of September, October and November. Several tons of topsoil had been lost due to water erosion. Thus, degraded land posed a severe problem to agriculture. The top layer of soil has vital components since all the nutrients required by plants had been present in this layer. This topsoil was, therefore, said to be the feeding zone of plants. It is the most valuable natural resource and it usually lies at some depth (of 15-20 cm) over the face of the land. We know that soil is not dead inert matter. It takes several years for an inch of topsoil to build up. Soil is also alive and dynamic, consisting of microorganisms such as bacteria, fungi, algae, protozoa, worms and insects but was lost much during these floods in Tamil country.

Peasants reclaimed the flooded land and began to cultivate again. Rulers granted exemption from payment of taxes to the flood-hit land of the farmers. European travellers, missionaries and East India Company officials reported the flooding in Tranquebar, Thanjavur, Cuddalore, Pondicherry and Madras. The French colonial government appointed a flood relief committee and made efforts to monitor the relief activities and it provided help in times of distress. It is found that many private individuals in Pondicherry generously contributed funds and helped the flood-affected population. Tsunamis and earthquakes that hit Tamil country have been reported but the extent to which it devastated the region has been not elaborated and details are found missing in the records. The simple reason was that the study of seismology had just begun at that time around the world.

NOTES

1. *Annual Report on Epigraphy* [henceforth *ARE*] (including Indian and South Indian Epigraphy published by the Government of India from 1881);

ARE, 64 of 1892; 375 of 1903; 158 of 1937-8; 125 of 1938-9; 308 of 1935-6.

2. *ARE*, 108, 112, 113, 118, 119, 121, 130, 131 of 1947-8, *ARE*, 8, 10 of 1948-9; *ARE*, 123 of 1938-9.
3. *ARE*, 38 of 1948-9.
4. Noboru Karashima, 'Allur and Isanamangalam: Two South Indian Villages of the Cola Times', in *Proceedings of the First International Conference-Seminar of Tamil Studies,* Kuala Lumpur, 1966, pp. 426-36; R. Tirumalai, 'Allur and Isanamangalam Revisited', in *Svasti Sri: Dr. Chhabra Felicitation Volume*, Delhi: Agam Kala Prakashan, 1984, pp. 25-55; R. Tirumalai, 'Land Reclamation of Flood Damaged and Sand-Cast Lands: A Study in Price, Rentals and Wages in Later Chola Times', in *Journal of the Epigraphical Society of India*, vol. I, 1984, pp. 64-7, see p. 66.
5. *ARE*, 113 of 1938-9.
6. *ARE*, 322 of 1906, 428 of 1905.
7. Archaeological Survey of India, *South Indian Inscriptions* [henceforth *SII*] Madras & New Delhi: Archaeological Survey of India, 1890 to 1990; *SII*, vol. VII, no. 96; *SII*, vol. VII, no. 97; *ARE*, 30 of 1903.
8. *ARE*, 87 of 1900; 88 of 1900.
9. *SII*, vol. VII, no. 193; *ARE*, 183 of 1901; *Epigraphia Indica* [hereafter EI], Calcutta & Delhi, XXII vols., Calcutta & Delhi, 1892-1978, vol. VII, p. 6.
10. *ARE*, 122 of 1937.
11. G. Vijayavenugopal, ed., *Pondicherry Inscriptions*, Pondicherry: French Institute, 2010, 157.
12. *SII*, vol. XXIV, nos. 65, 72, 85, 87, 89, 111, 113.
13. *ARE*, 87 of 1900.
14. *SII*, vol. XIV, no. 231.
15. *SII*, vol. XXIV, no. 23.
16. *ARE*, 422 of 1912.
17. Alfred Martineau, ed., *Memoires de Francois Martin, Fondateur de Pondichery, 1665-94*, 3 vols., Paris: E. Leroux, 1932-4, vol. II, pp. 176-7.
18. J. Talboys Wheeler, *Early Records of British India: A History of the English Settlements in India*, Calcutta, 1879, p. 80; Henry Davidson Love, *Vestiges of Old Madras: Traced from the East India Company's Records Preserved at Fort St. George and the India Office and from Other Sources*, 3 vols, London: 1913, reprint, Delhi: Asian Educational Services,1996, vol. II, p. 78; vol. III, p. 25, 217.
19. Aniruddha Ray, *The Merchant and the State: The French in India, 1666-1739*, 2 vols., Delhi: Munshiram, 2004, vol. II, p. 722.
20. Alfred Martineau, 'Les cyclones â la cote Coromandel de 1681 â 1916', in

Revue Historique de l'Inde Française, vol. II, 1917, pp. 229-324, see pp. 232-3.

21. Gnanou Diagou, ed., *Prathiyegamana Ananda Ranga Pillai Avargalin Sostha Ligitha Thinappadi Seythikurippu*, 8 vols., Pondicherry: Government of Pondicherry, 1948-54, vol. I, p. 219.
22. *Notices of Madras and Cuddalore in the Last Century from the Journals and Letters of the Earlier Missionaries of the Society for the Propagation of Christian Knowledge*, London: Longman & Co, 1858, p. 175.
23. *Der Königl. Dänischen Missionarien aus Ost-Indien eingesandter ausführlichen Berichten, Von dem Werck ihrs Ams unter den Heyden, angerichteten Schulen und Gemeinen, ereigneten Hindernissen und schweren Umstanden; Beschaffenheit des Malabarischen Heydenthums, gepflogenen brieflicher Correspondentz und mundlchen Unterredungen mit selbigen heyden [. . .]*, 1710-72, Teil 1-9 [Continuationen 1-108], Halle: Waiserihaus, (hereafter HB-Hallesche Berichte), HB 47. Cont., 1455-98.
24. Hippocrates, 'Air, Water & Places', in G.E.R Lyold, ed., *Hippocratic Writings*, London: Allen, 1978, pp. 12-16, 148-69.
25. K.M. Venkatramayya, *Administration and Social Life under the Maratha Rulers of Thanjavur*, Thanjavur: Tamil University, 1984, pp. 377-8.
26. Alfred Martineau, 'Les cyclones', p. 240.
27. Ibid., p. 242.
28. Ibid.
29. *Bulletin des Actes Administratifs des Etablissements francais de l'Inde* [hereafter BAA], L'Imprimerie du Gouvernement, Pondicherry, 1828-66, Arrête 470, 14 November 1832.
30. Alfred Martineau, 'Les cyclones', p. 240.
31. NAIP, *Bulletin Officiel*, Dépêche ministérielle, no. 79, Letter from the Minister of the Navy and Colonies dated 4 January 1872.
32. NAIP, *Bulletin Officiel*, Dépêche ministérielle, no. 51, Letter dated 2 January 1885.
33. Alfred Martineau, 'Les cyclones', pp. 292-4.
34. Tamil Nadu State Archives [hereafter TNSA], *Revenue Department*, G.O. No. 103, dated 7 November 1864.
35. TNSA, *Public Department*, G.O. No. 1020, dated 17 December 1884.
36. TNSA, *Revenue Department*, G.O. No. 3956, dated 20 December 1884.
37. TNSA, *Revenue Department*, G.O. No. 282, dated 21 December 1898.
38. W. Chambers, 'Some Account of the Sculptures and Ruins of Mavalipuram, a Place of a Few Miles North of Madras and Known to Seamen by the Name of Seven *Pagodas*', in *Asiatic Research*, vol. 1, 1788, pp. 145-70, see p. 154.

39. Ibid., p. 154.
40. Richard D. Oldham, 'Note on the Earthquake of 31 December 1881', in *Records of the Geological Survey of India*, 17 (2), 1884, pp. 47-53; R.E. Rogers, *Memorandum on the Earthquake of the 31 December 1881 and the Great Sea-waves Resulting therefrom, as Shown on the Diagrams of the Tidal Observatories in the Bay of Bengal, In General Report on the Operations of the Survey of India 1881-2*, Calcutta: Government Printing Office, 1883, pp. 70-3.
41. M. Ortiz and R. Bilham, 'Source and Rupture Parameters of the 31 December 1881 Mw' 7.9 Car Nicobar Earthquake Estimated from Tsunamis Recorded in the Bay of Bengal', in *Journal of Geophysical Research*, 2003, pp. 108-221.
42. Nicolao Manucci, trans., William Irvine, *Storia do Mogur, 1653-1708*, London: John Murray, 1907, vol. IV, p. 248.
43. *Notices of Madras and Cuddalore*, p. 175.

CHAPTER 6

The Study of Temperature and Atmospheric Pressure: Technology Transfer from Europe to Tamil Coast, Eighteenth-Nineteenth Centuries

Europeans who arrived on the Tamil coast realized that climatic conditions here were totally different from that of their own countries in Europe. They observed the different kinds of clouds which appeared very peculiar to their eyes, such as the ones of dark circles with golden strips that indicated rainy weather, and clouds of light blue or dark blue besides the golden and silver-coloured lightning that tore the sky during rainfall. Both missionaries and Company servants noted that intense low pressure areas often concentrated in the atmosphere in the form of coils, so they named them cyclones. The strong winds were storms as they moved from place to place in the sea. Sometimes, cyclones died out at sea, but then again, there were times when they crossed over the land causing untold devastation. The changes in the atmosphere attracted their attention during the course of their maritime trade in the Tamil region. Consequently, the Europeans made wholehearted attempts to explore the nature and climate of this region.

EXPLAINING THE PHYSICAL PHENOMENA: OBSERVATIONAL SCIENCE AND THE EUROPEANS ON THE TAMIL COAST

Europeans in Tamil country showed a keen interest to learn Hindu cosmography. Emanuel de Veiga (AD 1549-1605), the Jesuit who was at Chandragiri, studied and wrote in detail explaining the idea of the 'Cosmic Egg' in his letter dated 18 September 1599.[1] He noted that according to Hindu mythology, the universe had the form of a *brahmanda* (cosmic egg). Within it, there was water in which a tortoise floated; upon the tortoise there was a serpent topped by eight elephants. They supported the disk-shaped earth with the golden Meru Mountain as its centre: the sun, moon, stars and planets revolved around it.[2] According to Tamil literature of the medieval period, the Bhakti poets used the golden Mount Meru as a poetic metaphor and the mythological view of the earth and universe had continued. The concept of the earth as a sphere was unknown to the people.

William Stevenson, the Protestant chaplain at Madras in AD 1719, opined that a growing understanding and knowledge of science would help many in the East to recognize the advanced intellectual skills and knowledge of the Europeans. He also said that it would help persuade more people to embrace Christianity. Stevenson expressed his understanding and perspective on European intellectual dominance as being superior to that of the local communities, particularly in some fields. He suggested that the Society for the Promotion of Christian Knowledge (SPCK) should consider sending books from London to Madras on the theory and practice of physics. He felt that this would help the cause of schools set up by Protestant missionaries in Tamil country because the natives did not have adequate knowledge of physics.[3]

It is true that in the beginning the missionaries did not learn much about the existing knowledge of physics that the Tamils had obtained through the ages. Hindu works on science had divided the material world into *pancha bhutas* (the five elements) which were the earth (smell), fire (vision), air (feeling), water (taste) and ether (sound). Brahminical orthodoxy had created a sharp divide between the mental and the physical world, and this had largely prevented people of science from going beyond passive observation and intuition to practical experimentation, active theorizing and quantification. Nevertheless, physicists of the Nyaya Vaisesikha School had presented a dualistic view where the observed matter was said to be atomic, whereas the observing mind and time and space in which the universe existed was said to be continuous. What was the constituent of matter? What caused things to move, others to flow, some to lose shape and remain misshapen, while others reverted to the original whole form and some categories refused to lose shape except by being broken. There was not much understanding of these issues, so no answer could be thought of and sought for. Physics was considered to be the study of matter and energy, and the relation between them. It was science in pure form, since matter and energy were the basic constituents of the natural world. Tamil civilization had maintained its own history of science, so it was difficult for the European missionaries and servants of the trading companies to discover select and present this knowledge to Europe. Physics in the west then referred to experimental physics which had emerged. It was not to be confused with the older sense of natural science in general. Physics dealt with universal laws and with properties common to all substances. The general theories inferred from experimentation and observation had become indispensable to physicists of the early modern age. In contrast, from the time of Aristotle until the seventeenth century, physics, in Europe, had not emphasized experiment and quantification.

INTRODUCTION OF THE THERMOMETER AND THE BAROMETER BY THE FRENCH IN PONDICHERRY, AD 1709

The barometer and the thermometer were invented in AD 1643 and their use had spread to many parts of Europe. On the basis of these instruments, weather diaries began to be widely maintained since 1690. The thermometer, as it was then construed, was not applied to read the exact proportion of heat. It was generally thought that equal divisions of its scale represented equal tensions of caloric, but this opinion was not founded on any fact.[4] Most scientists believed in the existence of an objective property called temperature and wanted to know how to measure its true values. Thermometry began with no firm principles regarding the choice of thermometric substances. A bewildering variety of substances were tried to find the right thermometric fluid from among them. These included mercury, ether, alcohol, atmospheric air, sulphuric acid, linseed oil, water, salt water, olive oil, petroleum and lumps of clay. Only three of these known fluids became significant contenders for the claim of indicating true temperatures: atmospheric air, mercury and ethyl alcohol, the latter often referred to as 'spirit of wine' or simply 'spirit'. Finally, in the 1840s, it was found that the air thermometer was the best standard.[5]

Fahrenheit, working in Amsterdam, had established the use of mercury during the 1710s; small, neat and reliable, his mercury thermometers gained currency in many parts of Europe, partly through the physicians trained in the Netherlands. Réaumur preferred to use spirit, and made spirit thermometers popular in France. Mercury came to be preferred elsewhere by most people, including Anders Celsius (AD 1701-44), the inventor of the centigrade scale. Initially, people assumed that the thermometric fluid they used expanded uniformly with increasing temperature. After observing the disagreement between different types of thermometers, people modified this assumption to the assertion that one or another fluid expanded uniformly while others did not. Jean-André De Luc's investigations resulted in the conclusion that mercury was the most satisfactory thermometric liquid.[6] Since the spirit thermometer had been discredited in terms of comparability (and in other ways), Henri Victor Regnault, the renowned French chemist and physicist, made experiments and his main concern was to evaluate the air thermometer and the mercury thermometer. With regard to comparability, mercury betrayed clear signs of trouble. Regnault found that there was no such thing as the mercury thermometer. Mercury thermometers, made with different types of glass, differed from each other even if calibrated to read the same at the fixed points. Some had analysed the readings of barometers taken at different heights and temperatures. Others did it through direct measurement by a pressure gauge and a thermometer enclosed in a heated volume of air. The results varied from an expansion of 1 part in 172 to 1 part in 235 per degree centigrade. Many precautions were taken to remove unwanted moisture, until the year 1802, when Joseph Louis Gay-Lussac, the French chemist and physicist, finally fixed and gave 1:267 degree centigrade, which became the definitive value of the age.

The use of the barometer and the thermometer had spread in many parts of the world. The French in Pondicherry had a barometer and conducted experiments with it. The experiments conducted related to observations in physics and mathematics, and these were printed in AD 1692. In AD 1709, the details were sent to France along with notes on the experiments made at Pondicherry.[7]

THE STUDY OF WEATHER AND CLIMATE IN MADRAS AND NAGORE BY WILLIAM ROXBURGH USING NAIRNE & BLUNT THERMOMETER AND JESSE RAMSDEN BAROMETER, AD 1776-80

William Roxburgh, the surgeon of the English Company in Madras, made observations of pressure, temperature, wind and state of the weather at Fort St. George from October 1776 until May 1778.[8] He gave tables with seven columns with captions indicating (1) date of recording started from 1 October 1776, (2) hour noon onwards (time of observation), (3) thermometer within (temperature reading made from the thermometer located within the building, (4) thermometer without (temperature reading made from the thermometer located outside, (5) barometer (pressure reading with a barometer), (6) winds (strength: speed; points: direction), and (7) state of the weather and conditions at Fort St. George (notes on general weather patterns, i.e. cloudy, some rain in the night, etc.).[9] Roxburgh elaborated the way in which he did his meteorological observations. He said that a thermometer was kept without doors, and a barometer and thermometer kept within doors close together, for the sake of correcting the barometer if required. He observed them thrice a day. He also set down the direction and strength of the wind and the state of the weather. He distinguished four degrees of strength of the wind, namely, gentle, brisk, stormy and what was called a 'tufoon' in Tamil country, which he marked with the numbers, 1, 2, 3 and 4, besides no sensible wind, which was marked with a cipher. The state of the barometer was very uniform, varying only between 29.13 and 30.04 inches. A table was presented in six columns with notes made at the end of the months of October, November and

December of 1776, and of January and February 1777.[11] A table containing the greatest, least and mean heights of the thermometer and barometer in each month from March 1777 to May 1778 was also furnished.[12]

After Roxburgh's transfer from Madras to Nagore (near Nagapattinam), he continued to record the meteorological details at Nagore between 1778 and 1780.[13] He obtained a Nairne and Blunt thermometer from London for this purpose.[14] He also obtained a Jesse Ramsden barometer made by reputed scientific instrument makers in Britain and used it for his meteorological study.[15]

THE STUDY OF WEATHER AND CLIMATE BY COLONEL JAMES CAPPER IN MADRAS BY USING BAROMETER, AD 1777-8

Europeans began to use imported instruments and Colonel James Capper, the Comptroller-General of the Army and Fortification Accounts on the Coromandel Coast, conducted daily observations of the barometric pressure at Fort St. George in Madras from March 1777 to May 1778.[16] He also calculated the highest and the lowest barometric readings. William Roxburgh also gave the barometric pressure data for the same period, which does not tally with that of James Capper. The simple reason for this is the selection of two different locations by James Capper and William Roxburgh in Madras. Capper was posted at St. Thomas Mount in the interior while Roxburgh was posted at Fort St. George on the sea coast.[17]

JOHN GOLDINGHAM'S METEOROLOGICAL STUDIES AT THE MADRAS OBSERVATORY, AD 1796-1826

John Goldingham, employed at the Madras Observatory, mentioned that the barometer and the other instruments used, were placed two miles and three-quarters distance from the sea, and about 20 ft above sea level. It was known that in settled weather in India, the barometer showed a continual, regular rise and fall in 24 hours, and that the times of its greatest and least heights were nearly the same on different days, but Goldingham did not recollect that any observations upon a very extended scale had been made with the view of exhibiting such changes. In order, therefore, to obtain the quantities and times of these diurnal variations, he had the higher reading of the barometer taken every hour during three days of each month in 1823. These observations were given in detail by him.[18]

The rise and fall of the barometer were uniform throughout the entire 24 hours of each day of the year. At about 10 a.m. it was at its greatest height. It then began to fall, and continued falling. After 5 p.m. it reached the lowest. It then commenced rising, and before 11 p.m., the reading attained its highest point; and was again the lowest at 4 a.m. Goldingham stated that any change of wind or weather, of course, broke this chain of regularity, more or less according to the suddenness or violence of the change, but as there were no great or sudden changes at Madras in the 24 hours during 1823, the regularity was not much broken, scarcely at all indeed in the forenoon, but more at other times.[19]

The barometer was generally higher at about 10 a.m. than at 11 p.m., the times mentioned above, when it reached its greatest height in 24 hours; and it was lower at 5 p.m. than at 4 a.m.[20] Its greatest density was reached at about 10 a.m., diminishing till 5 p.m., when it began to regain what it had lost; and continued advancing towards its first state until about 11 p.m., when it had nearly the same density as in the forenoon. The diminution of density again began as in the day; the atmosphere, however, was not affected in so great a degree as when the sun was above the horizon.

It seemed that the higher readings taken during the usual interval of observing meteorological instruments, generally between sunrise and 8 or 9 p.m., were not those best calculated for the purpose of finding the exact mean. In order to ascertain what corrections should be applied on this account, the thermometer and hygrometer were also observed, along with the barometer, every hour during the interval stated above; also the winds and weather, together with the phases of the moon for each month; the day of the nearest approach of the moon to the earth in the month; and the day of her greatest distance. These observations were also given in detail.[21]

Goldingham gave details in a table with the heights of the

barometer, thermometer and hygrometer, taken from the ordinary diary on the days when the observations in the former table were made (the 10th, 20th and 30th of each month) with the differences of the means of both sets of observations (these differences were contained in a supplementary table), the corrections sought, and were applied to the daily means of the heights of the diary, as usually kept (a second supplementary table contained the corrections for the monthly means) and found by taking the means of the daily corrections, and also applied to the monthly means of the diary.

He stated that Madras was upon an open coast, and the time of the high-water at the Syzigies appeared to be at six hours and four minutes, the ebb and flow about six hours each way, with a rise and fall of little more than three ft.

If the times of these tides of the atmosphere, as they might be termed, as shown by the barometer, varied daily, like the times of the tides of the ocean, we might consider the moon as mainly instrumental in producing them. But occurring at nearly the same hours every day of the year, whatever may be the phases of the moon, or its position in its orbit, it was contented that the planet had only ordinary influence.[22]

The weather diary of Madras in 1823 clearly shows that every hour the circumstances regarding the higher readings and variation of the barometer with the winds and weather, and the position of the moon in its orbit on days, the barometer was affected by the luminary. The greatest and least heights of the barometer on the same days, with the other circumstances related to the moon, winds, and weather, had been found within an interval of years from the common diary. Therefore, one must not be led to the conclusion, after a particular examination, that the moon had any material influence in the changes of the atmosphere. One should not also arrive at a conclusion that the moon's influence, like the sun's, was at any time considerable, as regards the ordinary changes and motions of the atmosphere. While the moon raised the waters of the ocean, it gave light to the earth. Changes took place in the atmosphere regularly at or near the same times every 24 hours, while the times of the flux and reflux of the ocean were daily changing.[23]

TABLE 6.1: THE MEAN ANNUAL HEIGHTS OF THE BAROMETER AND THERMOMETER (WITH THE GENERAL MEAN OF EACH) SHOWN AT MADRAS, AD 1796-1821

Year	Barometer in inches	Thermometer in degrees Fahrenheit
1796	29.981	80.780
1797	29.948	83.430
1798	29.974	81.744
1799	29.988	81.430
1800	29.966	81.680
1801	29.966	81.354
1802	29.968	82.938
1803	30.006	82.571
1804	30.007	83.580
1805	30.067	82.046
1806	29.966	81.705
1807	29.923	79.738
1813	29.831	82.055
1814	29.891	81.304
1815	29.919	81.313
1816	29.972	80.434
1817	29.999	81.059
1818	29.950	81.088
1819	29.949	81.596
1820	29.950	81.546
1821	30.035	82.330
General Mean	29.964	81.700

Source: John Goldingham, 'Results of Meteorological Inquiries, Made at Madras', in *Transactions of the Royal Asiatic Society of Great Britain and Ireland*, vol. 3, no. 3, 1834, Appendix no. III, p. xxvii.

The hygrometer came to be used in Europe, and Horace-Bénedict de Saussure, De Luc's most prominent rival, disputed with him over the construction of hygrometers due to the cause of the cold that prevailed on mountain tops; but when it came to thermometry, Saussure, who had been close to the spiritist, Micheli du Crest, followed De Luc. According to the records, a hygrometer imported by the English Company was used in the Madras Observatory. Lieutenant Henry Kater of H.M. 12th Regiment communicated

TABLE 6.2: THE MEAN MONTHLY HEIGHTS OF THE BAROMETER AND THERMOMETER AT MADRAS, 1796-1821

Year	Barometer in inches	Thermometer in degrees Fahrenheit
January	30.085	75.168
February	30.076	77.157
March	30.041	79.920
April	29.955	82.417
May	29.851	86.918
June	29.861	88.159
July	29.867	85.645
August	29.879	84.732
September	29.908	83.825
October	29.942	81.858
November	29.956	78.672
December	30.074	78.843
General Mean	29.958	81.693

Source: John Goldingham, 'Results of Meteorological Inquiries, Made at Madras', in *Transactions of the Royal Asiatic Society of Great Britain and Ireland*, vol. 3, no. 3, 1834, Appendix no. III, p. xxvii.

his observation on the bearded seeds of a wild grass called *panimooloo* in Tamil likely to answer for an instrument of hygrometer. He also stated *yerudoovaal pullu* in the Carnatic (*Andropogon contortum* of Linnaeus) that grew in abundance likely to answer for an instrument of hygrometer.[24] Therefore, he made experiments. He said that the beards of the *panimooloo* grass were found by him to be perfectly suitable for the hygrometer. He, therefore, constructed three instruments and recorded the mean of their reading.[25] Kater gave an account of his first experiments on the fibre or beards of the *panimooloo.*[26] He was of the opinion later that the *yerudoovaal pullu* was far superior to any other hygroscopic substance hitherto discovered.[27] In this manner, some officers of the English Company began to explore how local objects were used by the natives in their own scientific work. Whether these were truly accepted by the Europeans or rejected is a separate issue. In general, there was a complete abandonment of Tamil knowledge and skill and furthering

TABLE 6.3: THE MEAN MONTHLY DRY OF THE HYGROMETER AT MADRAS, 1819-23

Month	Dry	Month	Dry
January	13.0	July	28.6
February	17.5	August	18.8
March	17.5	September	15.5
April	18.0	October	17.6
May	20.9	November	7.9
June	28.9	December	18.2
Mean	18.5		

Source: John Goldingham, 'Results of Meteorological Inquiries, Made at Madras', in *Transactions of the Royal Asiatic Society of Great Britain and Ireland*, vol. 3, no. 3, 1834, Appendix no. III, p. xxvii.

of European/western ideologies as well as adopting and strengthening the ideas Tamils borrowed from the West. Goldingham examined and analysed the data of hygrometer and published his findings.

From an examination of the three tables given above, it can be found that the hottest day in Madras, by the mean of all the daily observations during 21 years, was 15 June, when the mean height of the thermometer in 24 hours was 89.19 degrees Fahrenheit; the mean varied in different years from 95.1 degrees Fahrenheit to 81.6 degrees of Fahrenheit. The coldest day was 9 January, with the mean height of the thermometer on that day being 74.59 degrees of Fahrenheit; the mean varied from 77.1 degrees of Fahrenheit to 71.7 degrees of Fahrenheit. About 20 March and 20 October the thermometer was at its mean height.

The mean height of the thermometer, as deduced from the mean monthly heights, was 81.7 degrees of Fahrenheit; the greatest extreme of heat was 104.5 degrees of Fahrenheit; the former occurred in 1815 at 2 p.m. on 19 May, with a hot land wind blowing; and the latter on 12 January 1819, at about sunrise. Such extremes, however, were rare, the thermometer at Madras being seldom higher than 98 degrees of Fahrenheit or lower than 67 degrees of Fahrenheit.[28]

The hottest time of the 24 hours at Madras, taking the mean of the 12 months was about three-quarters of an hour past noon; this varied at different times of the year, from 11 a.m. to 3 p.m. The coolest time during 24 hours was at about 4.30 a.m., and the

TABLE 6.4: MEAN HEIGHT OF THE THERMOMETER AND BAROMETER DURING THE NORTH-EAST MONSOONS AT MADRAS BETWEEN AD 1796 AND 1821

Year	Thermometer in degrees Fahrenheit	Barometer in inches
October	81.858	29.942
November	78.672	29.956
December	75.843	30.074
January	75.168	30.085
February	77.157	30.076
March	79 .920	30.04
Mean	78.103	30.029

Source: John Goldingham, 'Results of Meteorological Inquiries, Made at Madras', in *Transactions of the Royal Asiatic Society of Great Britain and Ireland*, vol. 3, no. 3, 1834, Appendix no. III, p. xxix.

TABLE 6.5: MEAN HEIGHT OF THE THERMOMETER AND BAROMETER DURING THE SOUTH-WEST MONSOONS BETWEEN AD 1796 AND 1821

Year	Thermometer in degrees Fahrenheit	Barometer in inches
April	82.417	29.955
May	86.918	29.851
June	88.159	29.861
July	85.645	29.867
August	84.732	29.879
September	83.825	29.908
Mean	85.283	29.887

Source: John Goldingham, 'Results of Meteorological Inquiries, Made at Madras', in *Transactions of the Royal Asiatic Society of Great Britain and Ireland*, vol. 3, no. 3, 1834, Appendix no. III, p. xxix.

thermometer was usually at the mean height a little after 7 p.m., and at about 9 a.m.[29] According to the mean of the hygrometer during 1823, the atmosphere at Madras was least moist at about 2 p.m.; was most moist at 5.45 a.m.; and was in a mean state at 9.15 p.m., and at 10.15 a.m.

The barometer reading was highest in Madras upon a mean of years from AD 1796 to 1821 on 3rd January, and lowest about the end of May, the range being from 30.194 to 29.834 inches, but in the storms that had occurred during these years, the barometer was depressed in an extraordinary degree at Madras.[30] The barometer was 0.142 inches higher, and the thermometer 7.18 degrees of Fahrenheit lower, in the north-east monsoon than in the south-west monsoon as gleaned from the details above.

READING AND RECORDING THE DAILY BAROMETRIC PRESSURE AT THE MADRAS OBSERVATORY AD 1796-1843

Temperature and its variations constituted an important element of the climate of Madras. Europeans in the Tamil coast noticed that during the first part of the year, January to June, the increase of temperature by solar activities was greater than the loss by radiation and other actions. Hence the temperature rose steadily with the increasing elevation of the sun. During the latter part of the year, the balance was in the reverse way and temperature steadily decreased from July to December.

At the Madras Observatory, Goldingam recorded the atmospheric pressure data from AD 1796 to 1807. He made barometric pressure observations at sunrise, 10 a.m., noon, 2 p.m. and at sunset. From these, he calculated the daily mean and monthly values of pressure, corrected for temperature using the attached thermometer. The register maintained by him at the Madras Observatory does not contain atmospheric pressure data from 1808-12, as during this period Goldingham was on leave in Europe. On his return, he again maintained the meteorological register at the Observatory from 1812-25. This constitutes the first long-term barometric pressure readings at Madras.[31] Goldingham also gave additional information on the concurrence and strength of storms during 1812-21, and these have provided useful information. We find that very low pressures were recorded in May and June 1813, October 1818 and May 1820.

The register maintained by Goldingham was continued by Thomas Glanville Taylor, his successor. Taylor retained the daily observation times set by Goldingham, continuing the practice of deriving monthly averages from observations at sunrise, 10 a.m., 12 noon, 2 p.m. and at sunset until the end of 1832. By then, Taylor realized the different timings he had noticed: the highest pressure for the day was recorded at 10 a.m. and the lowest occurred at 10 p.m. So, at the beginning of 1833, he added 10 p.m. to his hours of observations. From January 1833, until the end of June 1837, additional observations at 10 p.m. were included in the monthly averages given by Taylor.[32] In 1837, he realized that he had only been observing maximum pressure and not the minimum that occurred around 4 p.m. Consequently, he reassessed the observation times and concluded that 10 a.m., 4 p.m. and 10 p.m. were the most useful hours to make observations. Observations for the hours of sunrise and sunset were deemed worthless due to their ever-changing time of occurrence, and that the 2 p.m. readings were also deemed of little value. Therefore, from 1 July 1837 onwards, the barometric pressure observations were confined to the hours of 10 a.m., 4 p.m. and 10 p.m.[33] Thus, from July 1837 till the end of 1843, Taylor's averages of monthly pressure were deduced from readings at 10 a.m., 4 p.m. and 10 p.m. It is important to point out that mercury barometers were used with attached thermometers and were located at an elevation of 8 m above sea level at Fort St. George. Ten different barometers were in use during the period 1830 and 1842.[34] However, the measurements followed were uniform in nature: the atmospheric pressures were measured in units of inches of mercury and temperature in degrees of Fahrenheit.

Captain S.O.F. Ludlow, Taylor's successor, also continued to compile data about atmospheric pressure and maintained the register in the Madras Observatory. Ludlow's observations were made at 8 a.m. He raised the barometer cistern from 8-16 m above mean sea level. This might have affected the pressure-time series. From 1841 onwards, barometric pressure at Madras was recorded in a standard form. The data compilations of the English East India Company in the Madras Observatory have been used on a large scale.[35]

The barometer and thermometer observations were entered in pencil in the convenient reading books in Madras as they were made. The standard barometer by Newman, supplied by the Royal Society of London in 1841, and of the two thermometers, verified at Kew, were recommended for refraction reductions; the one used being either a north or south window according to the direction of the wind at the time of observation.[36] Then the observations entered in pencil were carried out in the folio daybook, the originals were bound for preservation in the observatory. As the tropical climate seemed incapable of ensuring long-time preservation, copies of the records were made for safe deposit in England so that they could be readily made available for the purpose of reference both in London as well as in Madras whenever required.[37]

The arrangement of the daybooks was as follows: with the left side having details to be recorded for the polar distance with 12 columns (for reference number, date and barometer, thermometer, name of the observer, deduced circle reading, refractions, apparent polar distances, reductions to 1st January, mean polar distance of stars, apparent polar distance of stars, correction to experiments). The right side of the daybook had the details of right ascensions with 17 columns (namely, reference number, name of the object, number of wires, estimated magnitudes, mean of wires, inclination correction, collimation correction, meridian correction, personal equation of the observer, sums of column 6, 7, 8 and 9, corrected clock time of transit, clock correction applied, apparent right ascension, reduction to 1st January calculated, mean right ascension of stay, apparent right ascension by ephemeris, correction to clock or ephe-meris).[38] Two natives were appointed as the first and second assistants in the observatory, and they carried out the readings and maintained the registers.

The English Company in Madras had set up a bungalow at Dodabetta, situated 8,640 ft above sea level in the Nilgiri Mountains for conducting meteorological observations during 1847-8.[39] Later, Lieutenant Sykes continued to carry out meteorological observations at Dodabetta.[40]

WILLIAM GILCHRIST IN MADRAS AND HIS INVENTION OF SELF-REGISTERING BAROMETER AS WELL AS METALLIC TUBE BAROMETER, 1837

The lack of proper instruments and sparse supplies from England for the study of weather and climate were perennial problems faced by the English in Madras, which resulted in many Europeans drawing on their practical knowledge and creative facilities to invent instruments. In 1837, William Gilchrist, employed in a medical establishment in Madras, developed a self-registering barometer as well as a metallic tube barometer. The barometer, he claimed, recorded pressure continuously and with scientific accuracy. Gilchrist's self-registering barometer was constructed by suspending a barometer tube from one end of the balance, an apparatus being connected with the other end to record oscillations occasioned by the varying weight of mercury in the tube or by the varying pressure of the atmosphere on top of the tube.[41] A year later, Gilchrist claimed to have simplified the apparatus.[42]

The metallic tube barometer, the second instrument developed by Gilchrist, was intended to remove a major complaint among meteorologists regarding barometers. The common barometer was liable to be damaged as a result of the entry of air, and there was a great danger of destroying the instrument when attempting to expel the air by the only efficient method, boiling the mercury in the tube. This problem was even more acute in India where, once the air had entered the tube, the instrument became useless and only the maker of the instrument in England could repair it. Gilchrist substituted iron for glass in the vulnerable part of the barometer.[43] In this way, some Englishmen in Madras found the place conducive for scientific hypotheses. The scientific temperament displayed by the servants of the East India Companies came of age towards the end of an era in which natural philosophy predominated in the academies and at the beginning of a new age of science which was less theoretical, more demonstrative and quantifiable, and highly indebted to the progress of inventions and a public culture of displaying knowledge.

The temperature and weather of the Tamil coast had been

described variously by several travellers. The reason may be that it had changed and varied owing to the long span of time. Thomas Bowrey the English traveller who visited Madras in AD 1669, gives the details of the monsoon and the climate. He mentions that in the months of May and June there was the benefit of the sea breeze. There had been fresh gales which was something sulphurous and this may be due to the wind, more to the heat of the sun.[44] It is reported that on 22 August 1676, tide in Madras rose and fell from 2.5–3 ft.[45] Joseph Collet the English Governor of Madras wrote to his brother in AD 1718 and described that the sea breezes of the port of Madras had given him renewed vigour and that he was in perfect health.[46] Thomas Herbert the English traveler remarked that at Nagapattinam the climate was hot and wholesome, also in regard to the winds. He said that in some months of the year the heat was high and unseasonable and thus people were afflicted with fever, fluxes and other diseases.[47]

REPORTING THE CLIMATES OF MADRAS, TIRUCHIRAPALLI, MADURAI AND COIMBATORE DURING 1877 AND 1889

Europeans began to study the climate of the plains and hill stations in Tamil country. It is mentioned that Madras, although situated on the sea coast of India, had a drier atmosphere and a somewhat higher temperature than Masulipatnam. Its mean temperature was 82 degrees of Fahrenheit; that of December and January 76 degrees, and that of June 88 degrees, respectively. The afternoon temperatures were lowest in December (83 degrees), but the early morning readings were somewhat lower in January, when they average 68 degrees. The lowest reached in the course of the year, since the thermometers had been properly exposed, had been between 57.6 degrees and 62 degrees. In the years before 1877, they were many degrees higher. The highest readings, which were recorded with the hot land winds generally in May, had been between 102 degrees and 113 degrees in different years. The daily range of the thermometer is much the same as at Masulipatnam. The mean annual rainfall was only a little below 50 inches. In the wettest year on record (1827), it amounted to 88.4 inches, and in the driest (1832) it was only 18.5 inches. From June to September, there was less rain in Madras than at Masulipatnam, the average of each of these three months not exceeding 4 or 5 inches. But that of October was nearly 11 inches, and that of November, which was the wettest month of the year, nearly 14 inches. There was usually a spell of heavy rain in the early part of December; it was in this and the two preceding months that little cyclones frequently formed off the coast, when they passed across the Carnatic bringing a deluge of rain. From January to the end of April, there was little rain, and in May only a few thunderstorms, unless, as in some instances, the Carnatic was visited by a cyclone from the bay.[48]

Tiruchirapalli had the same mean temperature as Madras, viz. 82 degrees, but while it was 1 or 2 degrees warmer from February to May, it was 1 or 2 degrees cooler from June to September. Even in December and January, the afternoon readings of the thermometer averaged 85 degrees and 87 degrees, but the nights were comparatively cool in this season, though the thermo-meter rarely sank below 60 degrees on the coldest nights in the year. May was the hottest month, when the mean temperature was 88 degrees; but although the average maximum of the afternoon was 102 degrees, no reading higher than 108 degrees has been recorded, at all during 1877-89 (the last 12 years). In June, the temperature fell with the setting in of the strong west wind through the Palghat gap, and this brought cloudy skies. However, the atmosphere remained dry and there was very little rainfall. It was very dry, indeed, at all times, except in the last three months of the year. The mean humidity was 63 per cent, and in April, the driest month, only 54 per cent. Even in June and July, it was only 57 per cent, but in November it rose to 76 per cent. The mean annual rainfall was 37 inches. At the end of the month of April and in May thunderstorms were frequent, and the average rainfall of the latter month was nearly 4 inches. June and July were again dry months, with, on an average, only six rainy days in the two months. But rain became more frequent in August and September, and reached its maximum in October, in which month there were about 11 rainy days. November and the first-half of December were also more or less rainy, but the rainfall was lighter than at Madras.[49]

Madurai was a warmer place, having the same mean temperature as Tiruchirapalli and Madras (82 degrees of Fahrenheit). In December and January, it was 77 degrees, and in April and May it was 86 degrees. On the coolest night of the year, it rarely fell to 60 degrees, and the mean minimum of the three months, December to February, was between 68 degrees and 70 degrees. In April and May, the afternoon maximum readings averaged 101 degrees, but the highest recorded temperatures of 1877–89 had not been higher than between 104 degrees and 107 degrees. The rainfall was much higher than that of Coimbatore or than that of the plains to the south-east, viz., 35 inches in the year; and of this, nearly 5 inches of rainfall in April and May, about 9 inches altogether in August and September, and 9 inches in October. Although the rainfall was thus 60 per cent heavier than at Coimbatore, it rained less frequently, the average number of rainy days being only 59 in the year.[50]

Coimbatore also had a very dry climate, but being situated at a high level in the neighbourhood of lofty hills and within the influence of the Palghat gap, it was by no means a very hot station for the latitude. The mean temperature was 78 degrees, and 4 degrees lower than that of Tiruchirapalli and Madras, and in no months of the year did it deviate more than 5 degrees above or 4 degrees below this mean. In December and January, it was 74 degrees, in April 83 degrees, and from June to September it fluctuated between 76 degrees and 78 degrees. But the daily happening during the day was as much as 27 degrees in February and March, and 17 degrees in the dampest months of the year. The lowest readings of the year had varied between 54 degrees and 61 degrees in different years, and the highest between 99 degrees and 104 degrees in the shade. The mean humidity was 66 per cent, and that of February, the driest month, 52 per cent. There was less cloud than at Tiruchirapalli, and the rainfall was much lighter, viz., a little above 21 inches in the year. Of this, showers in April and May contributed more than 4 inches. Then, from June to September, although there were 6-10 rainy days in the month,

TABLE 6.6: THE HIGHEST TEMPERATURE REGISTERED WITH THERMOMETER FOR EACH MONTH AT MADRAS IN 1885 AND THE HIGHEST TEMPERATURE IN THE DAY

Month	Temperature in the Sun	Excess
January	146	61
February	146	62
March	161	69
April	150	59
May	150	55
June	151	53
July	153	54
August	154	59
September	160	65
October	149	60
November	148	62
December	149	65

Source: Henry F. Blanford, *A Practical Guide to the Climates and Weathers of India, Ceylon and Burma, and the Storms of Indian Seas*, London: Macmillan, 1889, p. 3.

in October and November the total fall amounted to 9 or 10 inches, and after the middle of December there was little rain till the April showers. Although rainfall was only about 21 inches, it rained on an average 85 days in a year.[51]

THE PRODUCTION OF ANGLICIZED WEATHER AND CLIMATE: THE HILL STATIONS OF OOTACAMUND AND WELLINGTON

It is mentioned that beside one or two smaller settlements and numerous plantations of coffee and cinchona (chiefly on the western slopes), two sanitaria, one civil and one for military purposes, had been established on the Nilgiri hills, one of the loftiest of the hill groups that dotted the southern part of the peninsula. The first, Ootacamund, was the summer residence of the Madras government; the second, Wellington, formerly known by its native name, Jakatalla, was exclusively occupied by British troops, but there was, in its vicinity, the civil settlement of Coonoor, the climate of which was very similar.

The Nilgiris is located on a lofty mass of hills at the southern extremity of the Mysore tableland, where the Eastern and Western Ghats converged. Their summit was a grassy undulating and hilly plateau, consisting of about 20 miles across, and averaged between 6,000 and 7,500 ft above the sea level. On its western margin it was crowned by a somewhat higher ridge. The western face fell away abruptly towards the plains of Malabar, as a continuation of the Western Ghats of Wynaad and Mysore. About midway it was crossed from north to south by another and higher range of hills, the culminating point of Dodabetta, which is 8,640 ft above the sea. Ootacamund and Wellington were on the opposite sides of this latter range. The former had been to the west, while the latter on its eastern flank. Thus Ootacamund was more exposed to the westerly monsoon which blew in the summer months. Wellington received more rain in October, while the wind had changed to east and north-east.

One striking characteristic of the Ootacamund climate, due to its position low down in the tropics, was the comparative uniformity of its temperature throughout the year. While the mean temperature of the year was 55 degrees, that of May, the warmest month, was only 4 degrees above this mean, and that of January, the coldest, only 7 degrees below it. Even this difference was due much more to the variation of the night temperature than to that of the day, and the latter was affected more by the amount of cloud and rain at different seasons than by the sun's declination and the length of the day. The afternoon temperatures were lowest in July, but varied very little from that month to December, after which they rose steadily till April. From April to July, they fell 10 degrees. The night temperatures were lowest in January, and rose 18 degrees between that month and May, from which time they varied only 1 degree or 2 degrees till October. The highest thermometer reading recorded in 1880, the only year for which a trustworthy register was 77.3 degrees and the lowest 25.3 degrees. The daily range was very great in clear weather, as was usually the case for tablelands. In January, February, and March it varied, on an average, between 27 degrees and 31 degrees in the 24 hours, but only from 10-14 degrees in the cloudy months, June to October.

The atmosphere was not very dry, except in March, the only month of 1880 in which the mean humidity ranged below 50 per cent of saturation. From June to November inclusive, the average of any month fell little below 80 per cent, and in November it was 90 per cent. From June to October, the skies were very cloudy, but as the rainfall register of Wellington showed, 1880 was a remarkably wet year on the Nilgiris, and there was reason to believe that Ootacamund was in general a less damp and cloudy station than might be.

In 1857, a geological survey of the Nilgiri hills was carried out during the summer monsoon with no more interruption from bad weather than would have been experienced in an English climate. Notwithstanding the prevailing cloudiness at this season, the rainfall was by no means heavy. From May to August, the weather was showery, but the total amount of rain was only 4 or 5 inches in the month, and out-of-door life was as little checked by weather as in an ordinary English summer. The heaviest rainfall was in October.[52]

At a level 1,000 ft lower, and with an additional screen interposed between it and the westerly monsoon, in the Dodabetta range, which dominates it by 2,500 ft, Wellington was 6 degrees warmer than Ootacamund, and had less cloud and rain, while in some respects it enjoyed even greater equability of temperature. During the summer monsoon, it was frequently fine and clear to the east of Dodabetta, while completely overcast and raining at Ootacamund. But after October, the relations of the two stations to the rainy winds were reversed, and in the final months of the year Wellington was the rainier station of the two.

The mean temperature of Wellington, as deduced from between 1875 and 1889 (14 years' registers) was 61 degrees. That of May, the warmest month, was only 5 degrees higher; and that of January, the coldest, only 6 degrees lower. The average afternoon temperature in May was 76 degrees, and the highest readings recorded in each of 1883-9 (6 years) had varied only between 79.5 degrees and 80.7 degrees, or little more than one degree. The early morning temperature in January averaged 45 degrees, and the lowest had varied between 34.2 degrees and 37.5 degrees, or a little more than three degrees in different years. In the winter months, the range during the 24 hours was considerably less than at Ootacamund. It amounted to 24 degrees in February, and 21 degrees only in January and March; and from June to November, it did not exceed between 13 degrees and 15 degrees on an average.

The air was drier than at Ootacamund. The humidity was above 80 per cent only in October and November, but it was below 70 per cent only from February to May. There was also appreciably less cloud than at Ootacamund, but clear skies predominated only from January to April. The rainfall was moderate, about 48 inches only in the year. From April to August, it fell 3-4 inches in each month on an average and it rained about every second day. In October and November, it was about three times as heavy, but even then, there averaged 13 or 14 fine days in each month. In January and February, rain was rare, and in March and April it rained on an average one day in five, chiefly in afternoon thunder storms.[53]

The English documents mention the ways in which hill stations and hill settlements became part of the annual migration of Britons

TABLE 6.7: MEAN DAILY RANGE OF THE THERMOMETER IN OOTACAMUND, WELLINGTON AND COIMBATORE IN 1885

Month	Ootacamund	Wellington	Coimbatore
January	31	21	23
February	29	24	27
March	27	21	27
April	21	20	24
May	18	17	21
June	12	14	18
July	10	13	18
August	12	14	17
September	13	15	19
October	14	13	17
November	16	13	17
December	22	17	19

Source: Henry F. Blanford, *A Practical Guide to the Climates and Weathers of India, Ceylon and Burma, and the Storms of Indian Seas*, London: Macmillan, 1889, p. 10.

living in Madras. The Nilgiris' climate was not only a relief from that of the Indian plains but it surpassed even that of England. The Nilgiris had been represented as paradise: as England on a grander scale, a land blessed with immortality. In the Nilgiris, the Englishman escaped from the tyranny of the regulated, ordered temporality into a space that stood outside of time. Nilgiris' nature was unburdened by social functions; its manifestations were fairylike, full of flowers and birds.[54]

CONCLUSION

There have been many changes in temperature, precipitation and other climatic elements through the periods in history. It may be mentioned here that the barometer and the thermometer were invented in AD 1643 and the use of rain gauges was made available in 1676 and it had spread in many parts of Europe only in 1690. On the basis of these instruments, weather diaries began to be maintained in Europe. Meteorological observations began in AD 1776 by William Roxburgh in Madras and he took the necessary measurements thrice a day using a Ramsden barometer and a Nairne thermometer. Roxburgh, who lived at Nagore near Nagapattinam from 1778 and 1780, studied the weather there during this period. He took systematic metrological records and his analysis was published in 1778 and 1790. It may be concluded that Tamil country witnessed remarkable climatic instability. The English had reported the climate of towns situated in the plains, such as Madras, Tiruchirapalli, Madurai and Coimbatore, during 1877 and 1889. They also wrote on the weather in England on comparison and the climate conducive to them at the hill stations of Ootacamund and Wellington.

NOTES

1. British Library [hereafter BL], London, MSS Sloane, no. 2748A, fols. 41r–45r; Sommervogel de Backer, *Bibliothèque des Ecriveins de la Compagnie de Jésus*, Paris: Oscar Schepens, 1810, vol. VI, p. 757; Jarl Charpentier, 'A

Treatise on Hindu Cosmography from the Seventeenth Century', in *Bulletin of the School of Oriental Studies*, vol. 3, no. 2, 1924, pp. 317-42, see p. 321.

2. Ibid.
3. See the letter of William Stevenson at Morningthrop in Newfolk, dated 8 December 1719, in A.W. Boehm, trans., in *Several Letters Relating to the Protestant Danish Mission in Tranquebar in the East Indies,* London: H. Clements, 1810, p. 49.
4. Joseph Louis Gay-Lussac, 'Enquiries Concerning the Dilatation of the Gases and Vapors', in *Journal of Natural Philosophy, Chemistry, and the Arts,* vol. III, 1802, pp. 207-16, 257-67, see p. 208.
5. Hasok Chang, 'Spirit, Air, and Quicksilver: The Search for the "Real" Scale of Temperature', *in Historical Studies in the Physical and Biological Sciences,* vol. 31, no. 2, 2001, pp. 249-84.
6. Jean-André De Luc, *Recherches sur les modifications de l'atmosphère,* 2 vols., Geneva, 1772, vol. I, pp. 275-83, 290.
7. 'A Comparison of the Barometrical Observations made in Different Places by M. Maraldi on 20 July 1709 in the Philosophical Memoirs of the Royal Academy of Sciences at Paris in 1709', in *The Philosophical History and Memoirs of the Royal Academy of Sciences at Paris: or An Abridgment of all the Papers relating to Natural Philosophy which have been Published by the Members of that Illustrious Society from the Year 1699 to 1720,* vol. III, London: Royal Society, MDCCXLII, pp. 238-9.
8. William Roxburgh, 'From a Meteorological Diary Kept at Fort St. George in the East Indies by William Roxburgh (1778) Communicated by Sir John Pingle' (Read on 29 January 1777), in *Philosophical Transactions of the Royal Society* [hereafter *Philosophical Transactions*], London: Royal Society, 1778, vol. 68, pp. 180-93, see p. 180.
9. Ibid., pp. 182-92.
10. *Philosophical Transactions,* vol. 70, p. 322.
11. *Philosophical Transactions,* vol. 68, p. 193. Under the column captioned 1776/77, Roxburgh listed 26 illnesses such as fevers, liver, liver cough, liver flux, epilepsy and fistula with month names, either single or double-digit numbers exist, referring to the number of patients. No legend or explanation is given in the article linking this table to the weather data reported.
12. 'A Table of the Greatest, Least and Mean Heights of the Thermometer and Barometer in Each Month from March 1777 to May 1778', in *Philosophical Transactions,* vol. LXX, p. 268.
13. William Roxburgh, 'A Continuation of the Meteorological Diary Kept at

Fort St. George in the East Indies, on the Coast of Coromandel by William Roxburgh, Assistant Surgeon in the Hospital at the Said Fort Communicated by Joseph Banks' (1 January 1780), in *Philosophical Transactions*, 1780, vol. LXX, pp. 246-71. On European attitudes towards India's climate and imperial expansion, see Mark Harrison, *Climates and Constitutions: Health, Race, Environment and British Imperialism in India, 1600-1850*, Delhi: Oxford University Press, 1999.

14. Edward Nairne was born in 1726 and apprenticed to one Matthew Loft at the age of 15, made Fellow of the Royal Society in AD 1776 and in 1797 was Master of the Spectacle Maker's Company. In 1794, Nairne opened his own workshop at no. 20, Cornhill and traded under his name. He turned out a great variety of apparatuses: barometers, air pumps, levels, manometers, thermometers, electrical machines and small portable equatorial based on J. Short's instrument.

15. Jesse Ramsden was a nephew of Abraham Sharp and started life as a clerk in a cloth warehouse before becoming apprenticed to Burton, an instrument maker in Denmark Street, Strand. In AD 1762, he began on his own in Haymarket and three years later married John Dolland's daughter, Sarah, receiving as part of the marriage settlement a share in the patent for manufacturing achromatic lenses. Ramsden first turned his attention to the improvement of J. Short's portable equatorial mounting. This he provided for the first time, with circles and counterpoises of correct size and weight and by 1773 he made at least four improved instruments. The telescopes were triple achromatic of two inches aperture and 15 inches focal length. He made a superior dividing machine in 1775 and a sextant graduated by it. Ramsden published the description of an engine for dividing straight lines on mathematical instruments in 1779. He sent Joseph Banks a paper entitled 'A Description of Two New Micrometers', another paper 'A Description of a New Construction of Eye-Glasses for such Telescopes as may be Applied to Mathematical Instruments'. He was one of the first to apply a positive compound eye piece to the microscope as a means of reading circle graduations. He died on 5 November 1800. Ramsden's larger instrument was the theodolite which he made in 1791. He made new instruments of the sextant and theodolite, devising new methods for their mounting, counterpoising, illumination and testing. Besides single telescopes fitted with acromatic object-glasses of his own manufacture, he made improved barometers, pyrometers, standard lengths, surveying chains, balances, electrical machines, levels, manometers and drawing instruments. See *Philosophical Transactions*, 1779, vol. LXIX, pp. 419-31; 1782, vol. LXXIII, p. 94; 1790, vol. LXXX, p. 134.

16. James Capper, *Observations of Winds and Monsoons*, Wittingham, London: Longmans, 1801, pp. 1-234.
17. R.J. Allan's suggestion that Capper's records were written in England after he had finished his duty in India and attributing to it the variance which occurred with that of Roxburgh's observations may be called into question. See Rob J. Allan, C.J. Reason, P. Carroll and P.D. Jones, 'A Reconstruction of Madras (Chennai) Mean Sea-level Pressure Using Instrumental Records from the Late Eighteenth and Early Nineteenth Centuries', in *International Journal of Climatology*, vol. 22, 2002, pp. 1119-42, see p. 1120.
18. John Goldingham, 'Results of Meteorological Inquiries Made at Madras', in *Transactions of the Royal Asiatic Society of Great Britain and Ireland*, vol. 3, no. 3, 1834, Appendix No. III, pp. xvii-xxiv, see p. xvii.
19. Ibid.
20. Ibid., p. xxii.
21. Ibid.
22. Ibid.
23. Ibid., p. xxiii.
24. 'An Account of Experiments made in Mysore Country in the Year 1804 to Investigate the Effects of Terrestrial Refraction by Lt. John Warren', in *Asiatic Researches Comprising History and Antiquities, the Arts, Sciences and Literatures of Asia* [hereafter *Asiatic Researches*], 20 vols., vol. IX, 1811, pp. 1-15; London: Murray, 1788–1839, reprinted, 24 vols., New Delhi: Cosmo, 1979.
25. Ibid., p. 6.
26. Ibid., pp. 16-23.
27. 'Description of a Very Serviceable Hygrometer by Lt. Henry Kater', *Asiatic Researches*, vol. IX, pp. 394-7, see p. 397.
28. John Goldingham, 'Results of Meteorological Inquiries Made at Madras', p. xxvii.
29. Ibid., p. xxviii.
30. Ibid., p. xxix.
31. BL, Oriental and India Office Collections [hereafter OIOC], J. Goldingham, *Tables Containing the Results of the Meteorological Observations taken at the Madras Observatory, 1796-1825*, (in microfilm), 1826.
32. BL, OIOC, Thomas Glanville Taylor, *Meteorological Register Kept at the Honorable the British East India Companies' Observatory at Madras for the Year 1822-43*, Madras Government, 1844 (in microfilm). See also T.G. Taylor, *Meteorological Register Kept at the Madras Observatory for August 1833*, Madras: Vepery Mission Press, 1834, vol. 1, p. 39; see 'Meteorological Register Kept at the Madras Observatory for the months of October 1833

to December 1834', 'Meteorological Register Kept at the Madras Observatory for the month of December 1833, January and February 1834', 'Meteorological Register Kept at the Madras Observatory for the month of March, April and May 1834', 'Meteorological Register Kept at the Madras Observatory for the month of June, July and August 1834', in *The Journal of Literature and Science*, published under the auspices of *Madras Literary Society and Auxiliary of the Royal Asiatic Society*, ed., J.C. Morris, FRS, vol. 1, 1834, pp. 199-205, 297, 368. It is noted that standard barometer no. 3 by Gilbert and standard thermometer by Troughton were used during 1833-4.

33. Ibid.
34. BL, European Manuscripts G 51, *Madras Observatory: Observations and Computations dated 1831–43 made at the Madras Observatory under Thomas Glanville Taylor.*
35. Rob J. Allan, C.J. Reason, P. Carroll and P.D. Jones, 'A Reconstruction of Madras (Chennai) Mean Sea-level Pressure Using Instrumental Records from the Late Eighteenth and Early Nineteenth Centuries, pp. 1119-42.
36. *Results of the Observations of the Fixed Stars made with the Meridian Circle at the Government Observatory, Madras in the years, 1862, 1863 and 1864 under the Direction of Norman Robert Pogson*, Madras: Government Press, 1887, p. xv.
37. Ibid., p. xvi.
38. Ibid.
39. *Meteorological Observations Made at the Meteorological Bungalow on Dodabetta 8640 Feet above the Level of Sea in the Years 1847-8*, Christian Knowledge Society & Male Asylum Press, Madras, 1848, p. 49.
40. Lieutenant W.H. Sykes, *Discussion of Meteorological Observations taken in India at Various Heights, Embracing those at Dodabetta on the Neelgherry Mountain*, London: Macmillan, 1850, p. 86.
41. William Gilchrist Esq, 'Description of a Plan for a Self Registering Barometer & Construction of Metallic Tube Barometer', *Indian Review and Journal of Foreign Science and Arts*, vol. II, 15 March 1837, pp. 16-19.
42. Ibid., pp. 17-18.
43. Ibid.
44. Thomas Bowrey, *A Geographical Account of the Countries Round the Bay of Bengal, 1669-79*, Cambridge: Cambridge University Press, 1905, p. 129.
45. BL, OIOC, Mss. *Original Correspondence*, no. 4215.
46. H. Dodwell, ed., *The Private Letter Books of Joseph Collet, Sometime Governor of Fort St George*, New York: Longmans Green & Co, 1933, p. 121 & 184.

47. Hippocrates, 'Air Water & Places' in G.E.R. Lyold, ed., *Hippocratic Writings*, London: Allen, 1978, pp. 148-69, 12–16.
48. Henry F. Blanford, *A Practical Guide to the Climates and Weathers of India, Ceylon and Burma, and the Storms of Indian Seas*, London, Macmillan, 1889, p. 181.
49. Ibid., pp. 181-2.
50. Ibid., pp. 182-3.
51. Ibid., p. 182
52. Ibid., pp. 119-22.
53. Ibid., pp. 122-3.
54. Kavita Philip, *Civilizing Natures: Race, Resources, and Modernity in Colonial South India*, New Brunswick: Rutgers University Press, 2004, p. 51-6.

CHAPTER 7

Concluding Remarks

Fernand Braudel (1902-85) of the Annales School laid out the environmental framework of Mediterranean history and he included the natural environment along with long-lived economic and cultural structures. It, thereafter, paved the birth of climate history.[1] Within his division of time—short, intermediate and long time—the natural hazards existed in short time. They were random and spectacular disruptions from the perspective of the people who experienced them. Their historical significance beyond the communities immediately impacted, was limited with a few notably destructive exceptions. Historical ecology, when connected to changes in human culture, took the shape of environmental history. Studies of change in past environments became essential to the evolution of environmental history. Emmanuel Le Roy Ladurie of the Annales School wrote on the medieval agrarian and environmental history of France and his work contained extremely rich weather data, poor harvests, famines, shortages and occasional years of abundance, and the way in which the difficulties were associated with the climatic stress of those times.[2] Christian Pfister, the geographer, wrote monographs about the early modern climate variability in Switzerland and the social and economic responses.[3] Pierre Alexandre, the climate historian, wrote about the climate history of Belgium and western Europe.[4] Thus, significant contributions had multiplied on the climatic history of Europe.

The historians and scientists worldwide have generally said that climate now and then had changed, besides the challenges ahead. They wrote about not only the ideas about the humans that usually sustained the discipline of history but also largely wrote the histories of pre-colonial, colonial, postcolonial, and post-imperial eras in response to the world wars and post-war scenario of decolonization and globalization. Some have called it the birth of the modern world but did not really outline the path and progress. In the regional histories of India, we find the history of weather and climate remains completely unchartered territory.

I

The rulers of Tamil kingdoms in the medieval period understood and managed the variability of the Indian monsoon. The greater depth and resolution at which historical events like rainfall on the Tamil coast is examined, the more it emerged that human contingency dominated the direct physical effects of weather and climate. Only by staying at the macroscopic level of generalization—at the synoptic scale—do the cause-effect propositions about climate and Tamil society make any sense. Despite theories suggesting increase in rainfall in Tamil region due to global warming, no significant trend has been observed confirmed by both parametric and non-parametric method. The significant trends, both increasing and decreasing, have been found at the Tamil regional level, suggesting slight change in distribution. Even though month-wise rainfall during monsoon has not changed, increase/decrease in different months has been noticed in some areas of Tamil country. Intensive and extensive agriculture had developed over the long span of time in Tamil country. Hence, attention to irrigation came to be paid and rainwater was stored being the only source.

II

It must be said that with the earliest descriptions, general accounts about the Tamil coast and the climate had revolved around two axes, the wet season (monsoon) and the dry. The relationship between climate and society is much more important and the historical knowledge found had been adequate in drawing firm conclusions about the nature of the Tamil country's climate.

Scholars have opined that some specific years stand out experiencing very severe El Nino (flow of ocean current) conditions in the world. The local, regional and global events of the years have been

TABLE 7.1: FAMINE YEARS WITH LESS ANNUAL RAINFALL (IN 10TH OF MM) BETWEEN 1871 AND 1900

Year	Rainfall
1871, Prajothpaththi	8934
1873, Srimukha	7850
1875, Yuva	7486
1876, Thaathu	5276
1878, Vehuthaanya	7750
1879, Pramaddhi	8545
1881, Vishu	7636
1885, Paarthibha	8496
1889, Virodhi	8107
1890, Vikrudhi	8385
1892, Nandana	7293
1894, Jaya	8250
1897, Yevilambi	8170
1899, Vikaari	7656
1900, Sarvaari	8979

Source: Regional Meteorological Centre, Chennai.

studied together to understand the impact of El Nino. The AD 1798 El Nino has been examined and it is said that this El Nino gave rise to severe drought conditions in Tamil region.[5] Further the drought of AD 1789-93 had been also compared closely with a strong El Nino event mentioned by William Roxburgh in Madras. Historically, El Nino events have been shown to coincide with low hurricane incidence. Therefore, in Madras, Roxburgh undertook a widespread programme of tree planting for the express purpose of encouraging above-average rainfall or of conserving existing rainfall patterns in Coromandel. El Nino's in 1720, 1728, 1828, 1878 and 1892 could be studied in a better way now, connecting the famines in Madras in 1719 and in 1728; the famine in Thanjavur in 1829; the famines in Madras and Madurai, both in 1878; and the famine in Madurai in 1892.

Famine outbreaks have been directly linked to a chain consisting of events like no rains, drought, bad harvest, food scarcity and escalation of prices. Rice being the staple food, its price in Tamil country had increased abnormally. It was sold at 175 rupees per *garce* in Madras during the famine of 1729-30. Again, we find 3-6 measures were sold for Re. 1 in Madras during the famine of AD 1781-2. Seven to eight measures were sold for Re. 1 in North Arcot region during the famine of 1804-7. Three to six measures of rice were sold for Re. 1 during the famine of 1823-4 in North Arcot. Thus, famines led to hunger, misery and sufferings and so people sold themselves, their infants and young children because of poverty and entered into slavery. Slave trade, therefore, flourished since it provided them the option of avoiding starvation and death.

III

Weather patterns in the Tamil coast had attracted a great deal of attention from colonial administrators, due in large part to the damaging effects of climate on European commerce and manpower. The cataloguing of instances of storms, hurricanes and cyclones on the Coromandel Coast was part and parcel of the over-riding disciplinary regime that was determined to make every corner of the French and British empires legible to the colonial gaze. While European accounts of storms had continued to appear since the seventeenth century, it was only from the beginning of the nineteenth century that colonial officers stepped up their efforts to catalogue and classify every aspect of storm occurrences and this was chiefly intended to simplify complex realities in order to make colonial statecraft possible. This process created elaborate artefacts of knowledge that necessarily privilege certain facts over others. This process is manifest in official narratives that framed storms as endangering the material instruments of the colonial apparatus, whatever other risks they have may posed to Tamil population. Some European Company officials and servants had in their narratives engaged with the European cooling climate of the early modern period. It must be mentioned that no natural law was derivable from the history of Coromandel cyclones that might allow us to predict the future. The only certainty was that such calamities will occur again and again and it had been impossible to draw any historical conclusion. Ananda Ranga Pillai, the chief dubash to the French Governor of Pondicherry, followed customary Hindu *shastric* sciences that he believed to be indeed capable of making such predictions. We may safely say that it is a simple narrative of man's struggle against the blind, insuperable power of nature.

IV

It is necessary to gain richer understandings of how the changing contours of climate like floods had contributed to changes in the physical and our society had responded and reacted. The floods damaged standing crops and thus peasants suffered largely.

TABLE 7.2: YEARS WITH HEAVY ANNUAL RAINFALL (IN 10TH OF MM) BETWEEN 1872 AND 1898

Year	Rainfall
1872, Angirasa	10675
1877, Isvara	10586
1884, Thaarana	11481
1887, Sarvajithu	11312
1893, Vijaya	10468
1898, Vilambi	11640

Source: Regional Meteorological Centre, Chennai.

The British and the French colonial governments in Madras and Pondicherry took steps to govern the people during floods which resulted from heavy rainfall. They began to re-evaluate the ways in which governments could better respond to flood disaster by appointing flood relief committees. The formation of the flood relief committee in Pondicherry with members of respected Indians in the formation and execution of emergency policy helped in the privatization of debris removal, and the ever-increasing use of internal public donation schemes and to raise flood relief funds.

V

A climate history of Tamil country for the period prior to the appearance of thermometers had not been conceivable. The meteorological details and the inconstancy of climate on timescales of decades and centuries mentioned in Tamil inscriptions cannot be of much use. For much of the medieval period, the climatology had been said to be rather dull. Owing mainly to the systematic work of the meteorological department, established by the Government of India in 1875, we have a far better knowledge of the weather and climate of Tamil country. Both meteorology and climatology had served to expand the field of science in the eighteenth and nineteenth centuries. The Europeans had observed very carefully the temperature and atmospheric pressure of the Coromandel Coast. They learnt through experience that the coast experienced the rainy season during the months of October and November, viz., retreating south-west monsoon period when there was a conflict between the waning south-westerly current and the advancing north-east monsoon winds. The direction of wind changed from north to north-east in December through east to south-west in June, thereby, giving rise to climatic changes. The general direction of wind changed to east in February and the temperature also increased. The wind, being feeble during the month, did not have the same soothing effect on the body against the high day temperatures as the sea breeze of the later months. The strength of wind increased steadily from March to June and its general direction changed to southerly shore winds in April and May. With the advent of European technology to the Tamil coast, instruments such as the thermometer, barometer, sympiesometer, anemometer and rain gauge came to be widely used at different places to conduct studies in meteorology and climatology.

VI

Scholars have begun to connect local data at the global level which, in my opinion, is an incorrect procedure and a historical blunder. The world cannot have a common and a single phenomenon at any point of time. The connectivity and occurrences of events must be looked at as largely probable ones. It is important to know the beliefs of the people regarding climate, particularly Hindus and the view on the natural disasters. In ancient India, it was believed that rainfall was caused by the sun (*Adityat Jayate Vrishti*). The idea that the forces of nature are at the service of divine command and that the disasters caused by natural hazards are the expression of a vengeful deities permeates Hindu perceptions of the universe. Such was the appeal of this image that it had merged with local belief systems to create syncretic explanations of these natural phenomena. On the one hand, it gave rise to notions of fatalism and un-worthiness among individuals and even entire communities regarded it as somehow the deserved product of their own actions. On the other, it led to a particular depiction of the Godhead as a stern and wrathful father.

During the age of European expansion on the Tamil coast in the early eighteenth century, the traders, missionaries and East India Company servants felt the urgent need to understand unfamiliar flora, fauna and geology both for commercial purposes and to counter environmental and health risks. Many East India Company physicians and surgeons entered into consulting service and they found employment with the French and British colonial governments acting as full-fledged professional and state scientists. Anxieties about environmental change, climatic change and extinctions of species and the fear of famine had helped to motivate early environmentalism. The climatic theorists of that period were armed with a conviction that change of climate might cause a transformation or even degeneration in man himself or the possibility of the disappearance of man. The apocalyptic environmental discourses of colonial scientists had frequently articulated a vision and a message of a far less cynical kind. The colonial environmentalists felt a steadily growing danger in which, they argued, the whole earth might be threatened by deforestation, famine extinctions and climatic change.

It must be stated that until the end of 1880s, a direct connection between decline in forest area and apparent regional increases in desiccation in India was not put forward as an argument for controlling deforestation. Surgeons began to define deforestation as a matter requiring serious policy initiatives and an expansion in the role of the British colonial state and their opinion had carried considerable weight. Further, individual conservation British propagandists like Hugh Clerghon and Robert Wright had often expressed a virulent hostility too much in the way of development of a colonial infrastructure. The threat posed by economic and social consequences of desiccation, first effectively promoted by Gibson Balfour and Cleghorn the English Company officials, had continued to preoccupy the official mind of the government throughout the period 1850-80. Moreover, a growing number of scientists in and outside medical service believed that they had found new evidence to support their theories linking deforestation with run-off, rainfall and famine incidence.[6] After the end of the company rule in 1857, the earlier established conservation policy was followed and continued and so we find the first Indian Forest Act being passed in 1865. However, the relation of forests to the hydrological cycle was also seriously thought by only some people.

In the studies of climate and history, we frequently find mention of climates 'worsening' and 'improving' and climates becoming more or less conducive for human development. European accounts note cold climate a good climate and the warm climate of Tamil country as bad climate and not conducive for their living. Such expressions need caution and they are also to some extent biased in the context of the use of the language, a phraseology which again shows improper judgements. Universalizing tendencies to moralize climate should not be allowed and so, the discourse about climate change or even the global climate cannot be reduced to a single index of 'global temperature' and so the modern concept of global warming becomes undesirable, if not dangerous. Scientists of the early twentieth century theorizing about carbondioxide increases and climate change, generally regarded such human-induced global warming from the historical perspective, may be important.

NOTES

1. Fernand Braudel, *La Méditerranée et le Monde Méditerranéen à l'époque de Philippe II*, Paris: Armand Colin, 1990; idem, *The Mediterranean and the Mediterranean World in the Age of Philip II*, Berkeley: California University Press, 1996.

2. Emmanuel Le Roy Ladurie, 'Histoire et Climat', in *Annales: économies, sociétés, civilisations*, vol. 14, 1959, pp. 3-34; idem, *Histoire du climat depuis l'an mil*, Paris, 1967; idem, *Times of Feast, Times of Famine: A History of Climate since the Year 1000*, trans., Barbara Bray, London: George Allen & Unwin, 1971.
3. Christian Pfister, *Agrarkonjunktur und Witterungsverlauf im Westlichen Schweizer Mittelland, 1755-97*, Bern: Geographistchen Institut der Universitat Bern 1975; idem, *Das Klima der Schweiz von 1525-1860 und seine Bedeutung in der Geschichte von Bevölkerung und Landwirtschaft*, Bern: Haupt Bern,1984; idem, *Wetternachhersage. 500 Jahre Klimavariationen und Naturkatastrophen (1496-1995)*, Bern: Haupt Bern, 1999.
4. Pierre Alexandre, *Le Climat au Moyen Âge en Belgique et dans les Régions Voisines*, Louvain: Centre Belge d'Histoire Rurale, 1976; idem, *Le Climat en Europe au Moyen Âge: Contribution à l'histoire des Variations Climatiques de 1000 à 1425, d'après les sources narratives de l'Europe occidentale*, Paris: Ecols des Hautes Etudes en Sciences Sociales, 1987.
5. Richard H. Grove, 'The Great El Nino of 1789-93 and its Global Consequences Reconstructing Extreme Climate Event in World Environmental History', in *The Medieval History Journal*, vol. X, 2007, pp. 75-98; Richard H. Grove and Vinita Damodaran, *Nature and the Orient, The Environmental History of South and South East Asia*, Delhi: Oxford University Press, 1998.
6. Richard H. Grove, *Green Imperialism: Colonial Expansion, Tropical Eden and the Origins of Environmentalism, 1600-1860*, Cambridge: Cambridge University Press, 1995, pp. 467-8.

APPENDICES

MONTHLY AND ANNUAL RAINFALL (IN 10TH OF MM) 1871-1900

Year	January	February	March	April	May	June	July	August	September	October	November	December	Annual
1871, Prajothpaththi	567	182	340	359	411	429	536	509	1,314	1,417	2,585	285	8,934
1872, Angirasa	15	84	108	537	1,077	575	777	1,018	1,176	1,425	3,010	873	10,675
1873, Srimukha	3	867	68	785	486	136	433	931	838	2,206	812	285	7850
1874, Bhava	0	190	27	360	1301	637	672	709	1,501	1,672	1,558	426	9053
1875, Yuva	37	42	137	540	621	520	350	1,154	815	2,146	800	324	7486
1876, Thaathu	6	0	179	305	725	324	586	843	673	435	771	430	5276
1877, Isvara	8	0	149	135	838	624	140	336	1,704	2,765	1,909	1978	10,586
1878, Vehuthaanya	206	0	4	726	869	616	636	1,356	1,129	936	932	340	7750
1879, Pramaathi	179	107	768	169	606	509	1,368	962	844	1675	858	500	8545
1880, Vikrama	112	96	28	440	547	250	389	1,251	595	2,009	3,393	966	10,076
1881, Vishu	151	1	87	20	716	259	319	1,401	1,312	866	1,623	881	7636
1882, Chitrabhanu	381	36	73	325	901	421	482	1,186	698	1,454	2,661	524	9,142
1883, Subhanu	16	363	158	107	922	524	535	970	588	2,470	1,848	899	9,400
1884, Thaarana	120	0	68	98	507	418	184	952	547	2,900	3,143	2,544	11,481
1885, Parthiba	41	0	162	61	501	688	311	427	1,124	1,834	1,981	1,366	8,496
1886, Viya	147	3	154	120	1,792	730	1,102	1,472	1,230	1,430	1,175	326	9,681
1887, Sarvajithu	1	52	272	420	552	779	412	1,093	844	2,992	2,127	1,768	11,312
1888, Sarvadaari	19	0	15	203	1,180	561	418	722	915	2,355	2,108	1,379	9,875
1889, Virodhi	64	8	163	629	521	349	1,167	1,149	1,279	1,484	583	712	8,107
1890, Vikrudhi	100	48	170	471	551	825	783	918	962	2,173	1,149	235	8,385
1891, Khara	99	531	195	398	332	385	170	533	695	3,406	1,177	1,151	9,072
1892, Nandana	21	164	5	576	428	902	728	1,963	643	1,288	248	328	7,293
1893, Vijaya	80	161	620	401	588	780	1,221	398	592	1,673	3,695	259	10,468
1894, Jaya	147	225	396	420	433	223	576	1,216	1,040	1712	1,361	501	8,250
1895, Manmatha	14	0	2	586	449	294	578	1,250	1,542	3,170	1,039	1,294	10,218
1896, Durmukhi	88	0	65	28	493	446	383	544	1,293	1,315	2,978	1,860	9,493
1897, Yevilambi	51	583	56	353	618	586	418	1,359	2,174	924	833	215	8,170
1898, Vilambi	132	135	5	552	663	301	392	889	1,793	2,396	2,995	1,387	11,640
1899, Vihaari	67	113	20	1,785	409	191	257	282	1,171	2,491	531	339	7,656
1900, Sarvaari	248	11	6	1,145	554	347	809	612	1,400	1,817	1,176	854	8,979

APPENDIX II

FAMINE YEARS (AS FOUND IN THE TAMIL CALENDAR AND IN THE CHRISTIAN CALENDAR)

Tamil Calendar Year	Christian Calendar Years (AD)
Prabhava	1687, 1747, 1807
Vibhava	1208
Sukila	–
Piramodhutha	1390, 1630, 1690
Prajothpaththi	1391, 1691, 1811, 1871
Angirasa	852, 1812
Srimukha	1153, 1813, 1873
Bhava	1814
Yuva	1215, 1635, 1875
Dhaathu	1396, 1876
Isvara	–
Vehudhaanya	1878
Pramaadhi	1879
Vikrama	1160, 1520, 1760
Vishu	1881
Chitrabhanu	–
Subhanu	1823
Thaarana	1644, 1824
Paarthiba	1765, 1825, 1885
Viya	1226, 1646
Sarvajithu	1647, 1827
Sarvadhari	1708
Virodhi	1829, 1889
Vikrudhi	1830, 1890
Kara	1531
Vijaya	1532, 1832, 1892
Vijaya	1833
Jaya	1054, 1894
Manmatha	1535
Durmukha	1536, 1776, 1836
Yevilambi	877, 1537, 1837, 1897
Vilambi	1538
Vihaari	1239, 1539, 1659, 1719, 1779, 1899

Contd.

APPENDIX II (*Contd.*)

Tamil Calendar Year	Christian Calendar Years (AD)
Saarvari	1540, 1900
Pilava	1241, 1661, 1781
Subakiridhu	1782
Sobakiridhu	1723, 1783
Krodhi	–
Visuvaavaha	1785
Parrabhava	1121
Pilavanga	1847
Keelaka	1188, 1788
Soumiya	1129, 1729
Saathaarana	1730, 1790
Virodhikiridhu	1131, 1791
Parithaabi	1792
Pramaadheesa	1853
Ananda	1734
Rakshasa	1855
Nala	–
Pingala	1737, 1797, 1857
Kaalayukthi	1798
Siddhaarthi	1019, 1620, 1799
Raudhra	–
Dunmathi	1201
Dundhubhi	1202
Rudhrodhkaari	1623, 1743
Rakthaashi	1804
Kurodhana	1805, 1865
Akshaya	1806, 1866

APPENDIX III

FLOOD YEARS (AS FOUND IN THE TAMIL CALENDAR AND IN THE CHRISTIAN CALENDAR)

Tamil Calendar Year	Christian Calendar Years (AD)
Prabhava	1867
Vibhava	–
Sukila	–
Piramodhutha	–
Prajothpaththi	–
Amgirasa	1812, 1872
Srimukha	–
Bhava	1754
Yuva	–
Dhaathu	–
Isvara	1157
Vehudhaanya	–
Pramaadhi	–
Vikrama	–
Vishu	–
Chitrabhanu	1402
Subhanu	–
Thaarana	1884
Paarthiba	–
Viya	1826
Sarvajithu	–
Sarvadhari	–
Virodhi	–
Vikrudhi	1710, 1830
Kara	1771, 1831
Vijaya	–
Vijaya	–
Jaya	–
Manmatha	1175, 1775
Durmukha	–
Yevilambi	937
Vilambi	1778, 1898

Contd.

APPENDIX III (*Contd.*)

Tamil Calendar Year	Christian Calendar Years (AD)
Vihaari	–
Saarvari	–
Pilava	–
Subakiridhu	1842
Sobakiridhu	1843
Krodhi	1004, 1784
Visuvaavaha	1125
Parrabhava	1126
Pilavanga	–
Keelaka	–
Soumiya	–
Saathaarana	1190
Virodhikiridhu	1071, 1191
Parithaabi	–
Pramaadheesa	–
Ananda	–
Rakshasa	–
Nala	–
Pingala	–
Kaalayukthi	–
Siddhaarthi	–
Raudhra	–
Dunmathi	1681
Dundhubhi	1682
Rudhrodhkaari	–
Rakthaashi	1864
Kurodhana	1745
Akshaya	–

Glossary

Aatru neer	River water
Aatrukulai	A tax collected on river-water use
Aayar	Herdsmen
Abishekham	Ritual ceremony
Anai odai	Dam passage
Anaikattu	Dam
Antarayam	Tax levied by local assembly
Arasaperru	Royal tax
Attrukaal vetti muttai al	Labour who executed river-water works
Attrupaattam	The lease of river-water
Avani	The month of August-September
Brahmanda	Cosmic egg
Candy	A unit of weight approximately 500 pounds
Chatty	A vessel in Tamil
Chettiars	Merchant caste
Dasavanda	A share consisting one–tenth of irrigation rights
Devadana	Temple lands
Dubash	Interpreter, Translator
Garce	A standard weight of rice a little less than four tons
Ghats	Mountain region
Gumastha	Clerk
Idaiyar	Shepherd community
Inamdar	Holder of *inam* land rights
Irai kaval	Local tax for police duty
Iraiyili	Tax free
Irandadi kinaru	Two feet depth well
Jala-sutrada	Water diviner
Kaaduvettis	Clearer of the forests
Kaal	Canal

Kacham	Resolution
Kadai madai	Last sluice
Kadamai	Tax paid to government on land
Kalam	A paddy measure, 1 *kalam* = 3 *tuni*
Kalani kulam	Tank in a paddy field
Kalanju	The weight of 72 grains
Kali yugam	The period of bad times
Kallanai	Dam built of stones
Kanikkai	Gift
Kanji	Rice gruel
Kanni kaal	Minor channel
Karai	Bund
Karpadi	Revetment
Karthigai	The month of November-December
Kartthumpu	Stone sluices
Kasu	A weight of metal, a copper coin
Kaviri karai thevai	Kaveri river-bank fund
Kilavan	Headman
Kinaru	Well
Kudimaraamath	The rendering of compulsory free labour by farmers and individual small proprietors
Kulam	Pond, artificial tank
Kumili	Sluice pit
Kurinji	Hilly region
Kuruni	A paddy measure, 1 *kuruni* = 8 *nali*
Kuttai	Small ponds
Kuzhi	A land measurement of 81 sq. ft
Maa	A land measure, 100 *kuzhis* = 1 *maa* of land
Madai	Paddle shutters
Mahasabha	The village assembly
Manigramam	Trade guild
Maravar	A caste
Marutham	Wetlands
Masi	The month of February-March
Masula	A kind of boat

Mathagu	Sluice
Meenavar	Fishermen
Mettu vaykaal	Upper-land channel
Mirasidar	The holder of *mirasi* land rights
Mullai	Forest region
Murai neer	Water system
Naaladi kinaru	Four feet depth well
Naali	A paddy measure, 1 *nali* = 2 uri
Nadu	A territorial division, a local assembly
Nattars	The officials of the Nadu
Nattu perumkal	Channel of the locality
Nayak	Chief or leader
Nayakathanam	Territory assigned by the Vijayanagara king to a *nayak*
Neer aarambam	The mouth of the water source
Neer amanji	Labour contributed for digging channel or desilting tank
Neer nilai	Water source
Neer odu kaal	Water passage, small water channel
Neer vilai	Cess levied on the use of water by the local assembly
Neer vilai kuli	Wages to be paid for the price of pumping water
Neerk kirai	Minor tax on water
Neerkiya vilai	Amount to be paid for pumping water
Neernilai kaasu	Tax payable for maintenance of the water-source in cash
Neithal	Coastal region
Nirkuli	Irrigation tax
Nulitteruvitta	String line
Olukkai	Branch channel
Ouragan	Hurricane
Paalai	Desert area
Pagoda	The gold coin called *varahan* with the image of Vishnu
Palams	A measurement of weight for silver, brass, copper and bronze

Panam	A gold coin used in Coromandel
Pancha bhutas	The five elements
Panchavara	A levy of one-fifth of the amount
Panimooloo	A wild grass in Tamil
Panjupili	Tax on ginned cotton
Parai kulam	Water pool for the Paraiyars
Parathavar	Pearl fisher
Pathakku	A unit of land measurement, 1 *pathakku* = 240 *kuzhis*
Peraaru	Major river
Periya kulam	Big pond
Periya vaykkal	Big channel
Peryeri	Big tank
Piastres	Silver dollars
Pon vari	Tax in money or gold
Ponnan madai	Paddle shutters named after the individual called Ponnan
Prasaatham	Sacred food offered to the deity
Pulattir kulam	Tank in a cultivation field
Punkunri madai	Paddle shutters named after the individual called Punkunri
Punya	Merit
Puratasi	The month of September-October
Puthu kinaru	New well
Real	The smallest Portuguese monetary unit
Sabha	Brahmin village assembly
Sabha viniyogam	Expenditure made by the *sabha*
Salangai	A type of boat
Samba	A variety of rice
Sammamdam	An agreement
Sandivighrahapperu	Ritual tax
Senneer amanji	Public service labour on the drinking water channel
Senneer poduirai	The common tax for good water
Senneer vetti	Free labour for rain water irrigation works
Sennir	Rain water

Senneer-vennir	Rain water-Waste water
Silasasanam	A written deed, agreement
Siru thumpu vaykkal	Channel from the minor river
Sitraaru	Minor river or Tributary
Sthanattar	Temple superintendent
Thalai madai	First sluice
Thinais	Landscapes
Thirumanjana kulam	Sacred tank for purification
Thonis	A kind of boat
Thotti	Cistern
Thumpu	Sluice, single valve system
Tuni	A measure, 1 *tuni* = 2 *pathakku*
Udichiru vaykkal	Inner minor channel
Ulliyak-kuli	Well-diggers wages
Ur kulam	Village tank
Uruni	Source of drinking water
Uruni kulam	Common tank for drinking water
Uruni naduvupatta kulam	Tank situated in the centre of the village
Uzhavar	Farmer
Valayar	The fishermen who used the net
Varigripakarana	Water works
Vath	Minor channel
Vay	Big water channel
Vaykkal	Water channel
Vedar	Hunter
Veli	A land measurement, 20 *ma* = 1 *veli*
Vellak kaal	Channel to the flood
Vellalan	Agriculturist
Velvi kumili	Outer sluice pit
Venneer	Tax towards the clearance of waste water
Vetti	Free labour for public works
Vettaparru	Tenure for the *vetti*
Vettipattam	A tax on pastoral community
Vittavan vilukaadu	Share holder's portion
Yeri	Lake
Yeri aayam	The cess collected on lake
Yeri meen kaasu	A tax levied and collected on the lake fish

Yeri meen paatam	Lease on lake fishermen
Yeri neer kovai	A tax levied on the use of lake water
Yeri patti	Maintenance of lake
Yeri vaariyam	Lake committee
Yeri vaay padikkaval	A tax collected towards the payment of salary to the person for police duty of the lake
Yeri vari	A tax on lake for its use
Yerudoovaal pullu	A grass grown in the Carnatic
Yettam	Piccottah
Zamindar	Landlord

Bibliography

PRIMARY SOURCES

I. Manuscripts

FRANCE

1. Archives de la Societe des Missions Etrangeres de Paris

Correspondance Pondichery -Paris

vol. 993, Lettre de Louis Mathon, 21 December 1774.
vol. 998, Lettre à M. Tesson, Karikal, 10 July 1833.
vol. 998, Lettre de Pierre Brigot à M. Tesson, 10 July 1833.
vol. 1004, Lettre de Mgr. Laouenan à MM. Les membres du Conseil de la Propagation de la Foi, Pondichéry, 18 December 1869.
vol. 1004, letter de Mgr. Laouenan à MM. Les membres du Conseil de la Propagation de la Foi, 4 November 1871.
vol. 1004, Lettre de M. Fourcade à M. Ligeon, Nangatur, 10 October 1874.
vol. 1004, Lettre circulaire de Mgr. Laouenan aux Missionnaires de son Vicariat, Pondichéry, April 1877.
vol. 1004, Lettre de M. Fourcade à Mgr. Laouenan du 17 April 1877 et 9 June 1877, Alladhy.
vol. 1004, Lettre de Mgr. Laouenan aux Missionnaires de son Vicariat, Pondichéry, 31 December 1877.
vol. 1004, Lettre de Mgr. Laouenan à Mgr. Hugarin, Evêque de Bayeux, Pondichéry, 28 June 1878.
vol. 1004, Circulaire de Mgr. Laouenan aux Missionnaires, Pondichéry, 8 July 1878.
vol. 1006, Lettre de Mgr. Chevalier aux Missionnaires, Bangalore, 6 February 1878.

2. Archives Nationales, Paris

MSS *Compagnie des Indes et Inde française*, Colonies Série C², vol. 196.

GERMANY

3. Archiv der Franckeschen Stiftungen, Halle

M2 B2:2, Wind and Weather Observations 1732-1737, Madras, 20 October 1732.
M2 B 2:14a, Meteorologische Observationen vom Jahr 1789.
M2 B 2:14b. Meteorologische Observationen vom Jahr 1790.
M2 B 2:17, Meteorologische Observationen vom Jahr 1791.
Wurzburger Geographische Arbeiten, vol. 80.

INDIA

4. National Archives of India, Regional Record Centre, Puducherry

Bulletin Officiel, Dépêche ministérielle, no. 79, Letter from the Minister of the Navy and Colonies dated 4 January 1872.
Bulletin Officiel, Dépêche ministérielle, no. 51, Letter dated 2 January 1885.
Proces-verbaux Conseil Colonial de Pondichery de 1877, Session Ordinaire, Séance du 28 Decembre 1877.

5. Tamil Nadu State Archives, Chennai

Public Department, G.O. no. 1020, dated 17 December 1884.
Revenue Department, G.O. no. 103, dated 7 November 1864.
Revenue Department, G.O. no. 3956, dated 20 December 1884.
Revenue Department, G.O. no. 282, dated 21 December 1898.

ITALY

6. Archivum Romanum Societatis Iesu, Rome

MSS, Goa 56, Letter of Fr Pierre Martin to Le Gobien in December 1703 in Lettera Annua della Missione del Madure, December 1703.

THE NETHERLANDS

7. Nationaal Archief, Den Haag

MS VOC 876, VOC 1082, VOC 1147, VOC 1156.

UNITED KINGDOM

8. British Library, London

Oriental and India Office Collection

Goldingham, John Results of the Meteorological Observations taken at the Madras Observatory, 1796-1825 (in microfilm).

L/PJ/6/107

MS Eur. G 51, *Madras Observatory: Observations and Computations dated 1831-43 made at the Madras Observatory under Thomas Glanville Taylor.*

MSS Sloane, no. 2748A

Original Correspondence, nos. 1748, 3247, 4215.

Taylor Thomas Glanville, *Meteorological Register Kept at the Honorable British East India Companies' Observatory at Madras for the Year 1822-43*, Madras Government, 1844 (in microfilm).

V/4/session 1867/volume 52, *Copies of Papers Relating to the Famine in Madras Presidency*, no. 35, proceedings of the Madras Government, Revenue Department, 25 July 1866, enclosed in GoM to SoS, dated 11 August 1866.

V/23/192, *Collection of Papers Relating to Control of Government over Water in the Madras Presidency, 1851-9.*

II. PRINTED DOCUMENTS AND CONTEMPORARY CHRONICLES

IN DUTCH

Coolhas, W.Ph., ed., *Generale Missiven van de Gouverneurs-Generaal en Raden Aan Heren XVII der Verenigde Oostindische Compagnie*, 7 vols., 'S Gravenhage, 1960-84.

Pieters, Sophia., *Memoir of Hendrick Zwaden 1697*, Colombo, Government Press, 1911.

IN ENGLISH

'A comparison of the barometrical observations made in different places by M. Maraldi on 20 July 1709 in the Philosophical Memoirs of the Royal Academy of Sciences at Paris in 1709', in *The Philosophical History and Memoirs of the Royal Academy of Sciences at Paris: or An Abridgment of all the Papers relating to Natural philosophy which have been published by the Members of that Illustrious Society from the Year 1699 to 1720*, vol. III, London, MDCCXLII, pp. 238-9.

'An Account of Experiments made in Mysore Country in the Year 1804 to Investigate the Effects of Terrestrial Refraction by Lt. John Warren', in *Asiatic Researches Comprising History and Antiquities, the Arts, Sciences and Literatures of Asia*, 20 vols., London: Murray, 1788-1839; vol. IX, 1811, pp. 1-15, reprinted, 24 vols., New Delhi: Cosmo, 1979.

'A Table of the Greatest, Least and Mean Heights of the Thermometer and Barometer in Each Month from March 1777 to May 1778', *Philosophical Transactions*, vol. LXX, p. 268.

Blanford, Henry F., *A Practical Guide to the Climates and Weathers of India, Ceylon and Burma, and the Storms of Indian Seas*, London, Macmillan, 1889.

Boehm, A.W., trans., *Several Letters Relating to the Protestant Danish Mission in Tranquebar in the East Indies*, London: Society for the Propagation of Christian Knowledge, 1810.

Capper, James, *Observations of Winds and Monsoons*, Wittingham, London: Longmans, 1801.

Chambers, W., 'Some Account of the Sculptures and Ruins of Mavalipuram, a Place of a Few Miles North of Madras and Known to Seamen by the Name of Seven Pagodas', in *Asiatic Research*, vol. 1, 1788, pp. 145-70.

Dalyell, R.A., *Memorandum on the Madras Famine of 1866*, Madras: Government Press, 1867.

'Description of a Very Serviceable Hygrometer by Lt. Henry Kater', in *Asiatic Researches*, vol. IX, pp. 394-7.

Dodwell, H., ed., *The Private Letter Books of Joseph Collet, Sometime Governor of Fort St George*, New York: Longmans, Green & Co, 1933.

Foster, William, *The English Factories in India: A Calendar of Documents in the India Office, British Museum and Public Record Office*, XIII vols., Oxford: Clarendon Press, 1906-17.

Franklin, J.J., 'Notice of the Storms by J.J. Franklin', in *Madras Journal of Literature and Science*, vol. XIV, 1847, pp. 146-51.

Gilchrist Esq, William, 'Description of a Plan for a Self Registering Barometer & Construction of Metallic Tube Barometer', in *Indian Review and Journal of Foreign Science and Arts*, vol. II, 15 March 1837, pp. 16-19.

Goldingham, John, 'Results of Meteorological Inquiries Made at Madras', *Transactions of the Royal Asiatic Society of Great Britain and Ireland*, vol. 3, no. 3, 1834, pp. i-xxx.

Gay-Lussac, Joseph Louis, 'Enquiries Concerning the Dilatation of the Gases and Vapors', in *Journal of Natural Philosophy, Chemistry, and the Arts*, vol. III, 1802, pp. 207-16, 257-67.

Halley, Edmund, 'An Historical Account of the Trade Winds, and Monsoons Observations in the Seas between the Tropics with an Attempt to Assign the Physical Cause of the Said Winds', in *Philosophical Transactions of the Royal Society of London*, vol. 16, 1686, pp. 153-68.

Mackenzie, A.T., *Official Papers Concerning the Construction of the Madras Harbour*, Madras: Government Press, 1902.

Meteorological Observations Made at the Meteorological Bungalow on Dodabetta 8640 Feet above the Level of Sea in the Years 1847-8, Madras: Christian Knowledge Society & Male Asylum Press, 1848.

'Meteorological Register Kept at the Madras Observatory for the Month of June, July and August 1834', in *The Journal of Literature and Science* (published under the auspices of *Madras Literary Society and Auxiliary of the Royal Asiatic Society*), ed., J.C. Morris, FRS, 1834, vol. 1, pp. 199-205, 297-368.

Naidu, Kurma Venkata, *The Revised Irrigation Bill: Being a Reprint of a Series of Articles Contributed to the Indian Patriot*, Madras: Loganadham Bros., n.d.

Notices of Madras and Cuddalore in the Last Century from the Journals and Letters of the Earlier Missionaries of the Society for the Propagation of Christian Knowledge, London: Longman & Co, 1858.

Oldham, Richard D., 'Note on the Earthquake of 31 December 1881', in *Records of the Geological Survey of India*, 17 (2), 1884, pp. 47-53.

Oldham, Thomas, 'A Catalogue of Indian Earthquakes', in *Memoir of Geological Survey of India*, vol. 19, 1883, pp. 163-215.

Records of Fort St. George: *Diary & Consultation Book, 1672 to 1756*, Madras: Government Press, 1910-50.

Report of the Indian Famine Commission, Part I: *Famine Relief*, London: Madras: Government Press, 1880.

Results of the Observations of the Fixed Stars Made with the Meridian Circle at the Government Observatory, Madras in the Years, 1862, 1863 and 1864 Under the Direction of Norman Robert Pogson, Madras: Government Press, 1887.

Rogers, R.E., *Memorandum on the Earthquake of the 31 December 1881 and the Great Sea-Waves Resulting Therefrom, as Shown on the Diagrams of the Tidal Observatories in the Bay of Bengal, In General Report on the Operations of the Survey of India 1881–1882*, Calcutta: Government Printing Office, 1883.

Roxburgh, William, 'From a Meteorological Diary Kept at Fort St. George in the East Indies by William Roxburgh (1778) Communicated by Sir John Pingle' (Read on 29 January 1777), in *Philosophical Transactions of the Royal Society*, London: Royal Society, vol. 68, 1778, pp. 180–93.

——, 'A Continuation of the Meteorological Diary Kept at Fort St. George in the East Indies, on the Coast of Coromandel by William Roxburgh,

Assistant Surgeon in the Hospital at the Said Fort Communicated by Joseph Banks' (1 January 1780), in *Philosophical Transactions*, vol. LXX, 1780, pp. 246-71.

Smith, R. Baird., 'Memoir on Indian Earthquakes, Part I', in *Journal of the Asiatic Society of Bengal*, vol. 11, 1843, pp. 1046-54; Part II, vol. 12 (2), New Series, 136, pp. 1029-56; Part III, vol. 156, 1884, pp. 964-83.

——, *The Cauvery, Kistnah, and Godavery: Being a Report on the Works Constructed on these Rivers for the Irrigation of the Provinces of Tanjore, Guntoor, Mausilpatam, and Rajahmundry in the Presidency of Madras*, London: Smith, Elder & Co., 1856.

Strachey, Richard, 'On the Alleged Correspondence of the Rainfall at Madras with the Sun-Spot Period and on the True Criterion of Periodicity in a Series of Variable Quantities', in *Proceedings of the Royal Society of London*, vol. 26, 1877, pp. 249-61.

Sykes, Lieutenant W.H., *Discussion of Meteorological Observations taken in India at Various Heights, Embracing those at Dodabetta on the Neelgherry Mountain*, London: Macmillan, 1850.

Taylor, Thomas Glanville, *Meteorological Register Kept at the Madras Observatory for August 1833*, Madras: Vepery Mission Press, 1834.

Voelecker, John Augustus, *Report on the Improvement of Agriculture*, London: Madras: Government Press, 1893.

IN FRENCH

Bertrand, Joseph, *La Mission du Madure d'apres des Documents Inedits*, 4 vols., Paris: Imprimerie de Poussielgue, 1847-54.

Bulletin des Actes Administratifs des Etablissements francais de l'Inde, L'Imprimerie du Gouvernement, Pondichery, 1828-66.

Gaudart, Edmond, and Alfred Martineau, eds., *Correspondance du Conseil Supérieur de Pondichéry avec le Conseil de Chandernagor du 30 Septembre 1728 au 10 Fevrier 1757*, 3 vols., Paris Pondichery: Societe de l'Histoire del'Inde Francaise, 1915–19.

——, and Alfred Martineau, eds., *Procès-Verbaux des Deliberations du Conseil Superieur de Pondichery du 1 er Fevrier 1701 au 31 Decembre 1739*, 3 vols., Pondichery: L'Imprimerie du Gouvernement, 1913-15.

Lettres Edifiantes et Curieuses Ecrites des Missions Etrangeres par Quelques Missionaries de la Compagnie de Jesus, 28 vols., Paris: Chez N.E. Sens, Chez J. Varnavel, 1780-94.

Martineau, Alfred., ed., *Memoires de Francois Martin, Fondateur de Pondichery, 1665-1694*, 3 vols., Paris: E. Leroux, 1932-4.

Revue Coloniale, tome XVIII. Paris: Imprimerie et Librarie Adminsitative de Paul Dupont, 1857.

IN GERMAN

Der Königl. Dänischen Missionarien aus Ost-Indien eingesandter ausführlichen Berichten, Von dem Werck ihrs Ams unter den Heyden, angerichteten Schulen und Gemeinen, ereigneten Hindernissen und schweren Umstanden; Beschaffenheit des Malabarischen Heydenthums, gepflogenen brieflicher Correspondentz und mundlchen Unterredungen mit selbigen heyden, Teil 1-9, (Continuationen 1–108), Halle: Waiserihaus, 1710-72.

Knapp, Georg Christian, et al., *Neuere Geschichte der evangelischen Missions-Anstalten zu Bekehrung der Heiden in Ostindien aus den eigenhändigen Aufsätzen und Briefen der Missionarien erausgegeben,* Teil 1-8, (Stück 1-95), Halle: Waisenhaus, 1770-8/1795-1848.

IN PORTUGUESE

Correia, Gaspar, *Lendas da India,* 4 vols., reprint, Porto 1975.

Stephen, S. Jeyaseela, *Letters of the Portuguese Jesuits from Tamil Countryside, 1666-88,* Pondicherry: Institute for Indo-European Studies, 2001.

IN TAMIL

Aiyar, U.V. Swaminatha, ed., *Perungathai,* Madras, Paari Nilayam, 1924.

——, *Puranaanuru,* Madras, 1971.

Alalasundaram, ed., *Ananda Ranga Pillai Naatkurippu,* vols. 9-12, Pondicherry, Self published, 2005.

Charier, C. Jegannatha, ed., *Erelupadu by Kambar,* Madras: The Author, 1967.

Diagou, Gnanou, ed., *Prathiyegamana Ananda Ranga Pillai Avargalin Sostha Ligitha Thinappadi Seythikurippu,* 8 vols., Pondicherry: Government of Pondicherry, 1948-54.

Gobalakichenane, *Irandaam Veera Nayakkar Naatkurippu, 1778-92,* Madras: Nattamil Pathippagam, 1992.

Irayanar Ahapporul: Traditional Commentary based on Nakkirar, Madras: C.B. Namasivaya Mudaliar, 1943.

Nattar, N.M. Venkatasamy, and R. Venkatachalam Pillai, eds., *Aganaanuru,* 3rd edn., Madras: South India Saiva Siddhantha Pathippu Kazhagam, 1957.

Nattar, Venkatasamy, ed., *Manimekalai*, Tirunelveli: South India Saiva Siddhanta Pathippu Kazhagam, 1992.

——, ed., *Silapathikaaram*, Tirunelveli: South India Saiva Siddhanta Pathippu Kazhagam, 1992.

Pillai, Avvai Duraisamy, ed., *Pathirrupathu*, Tirunelveli: South India Saiva Siddhanta Pathippu Kazhagam, 2nd edn., 1955.

Pope, G.U., *The Sacred Kural of Thiruvalluva Nayanar*, Oxford: Oxford University Press, 1886.

Price, Frederick, H. Dodwell, and V. Rangachari, trans & ed., *The Private Diary of Ananda Ranga Pillai*, 12 vols., reprint, Delhi: Asian Educational Services, 1980.

Stephen, S. Jeyaseela, *The Diary of Muthu Vijaya Thiruvengadam Pillai, 1794-96*, Pondicherry: Institute for Indo-European Studies, 1999.

——, *Rangappa Thiruvengadam Pillai Naatkurippu, 1760-6*, II vols. Pondicherry: Pondicherry Institute of Linguistics and Culture, 2000.

——, *The Diary of Rangappa Thiruvengadam Pillai, 1760-8: Translated from Original Tamil with Notes*, Pondicherry: Institute for Indo-European Studies, 2001.

——, *Arangappa Thiruvengadam Pillai Naatkurippu, from 13 June 1767 to 29 December 1769*, (Vol. III, in Tamil), Chennai: Sekhar Publishers, 2006.

Sundaram, P.S., trans. and ed., *The Kural*, New Delhi:Munshiram, 1990.

III. Travelogues

Bowrey, Thomas, *A Geographical Account of the Countries Round the Bay of Bengal, 1669-79*, Cambridge: Cambridge University Press, 1905.

Cortesao, Armando, trans. and ed., *The Suma Oriental of Tome Pires and the Book of Rodrigues*, 2 vols., reprint, New Delhi: Asian Educational Services, 1990.

Manucci, Nicolao, trans., William Irvine, *Storia do Mogur, 1653-1708*, 4 vols, London: John Murray, 1907.

IV. Epigraphical Sources

Aiyar, K.R. Srinivasa, *Inscriptions of Pudukottai State*, Pudukottai: Pudukottai State, 1929.

Archaeological Survey of India, *South Indian Inscriptions*, XXVI vols., Madras: Archaeological Survey of India, New Delhi: Government of India, 1890-1990.

Kamal, S.M., *Sethupathi Mannar Seppedugal*, Ramanathapuram: Sharmila Pathippagam, 1992.

Kasinathan, N., and R. Nagaswamy, eds., *Kanyakumari Kalvettugal*, 5 vols., Madras: State Department of Archaeology, 1968-79.

Rao, Gopinatha T.A., ed., *Travancore Archaeological Series*, 9 vols, Madras & Trivandrum: Travancore State, 1910-41.

Rao, Rama M., *Telugu Inscriptions of Andhra Desa*, Tirupati: Sri Venkateswara University, 1969.

Sadhu Subrahmanya Sastry, *Tirumalai-Tirupati Devasthanam Epigraphical Series*, 6 vols., Madras: Tirumalai-Tirupati Devasthanam, 1931-8.

Srinivasan P.R., and Marie Louise Reniche, *Tiruvannamali: Inscriptions*, Pondicherry: Ecole Francaise d'Extreme-Orient, 1990.

Subrahmanyam, T.N., *Pandiyar Seppedugal Pathuu*, Madras: Tamil Varalaatrru Kazhagam, 1967.

Subramanian, T.N., *South Indian Temple Inscriptions*, Madras, III vols, Madras: Government Oriental Manuscripts Library, 1953-7.

Vijayavenugopal, G., ed., *Pondicherry Inscriptions*, Pondicherry: French Institute, 2010.

SECONDARY SOURCES

ARTICLES

Ahuja, Ravi, 'State Formation and Famine Policy in Early Colonial South India', in Sanjay Subrahmanyam, ed., *Land, Politics and Trade in South Asia*, Delhi: Oxford University Press, 2004, pp.147-85.

Brentjes, Buchard, 'Meteorlogische beobachtungen des 18.Jahrhunderts in Madras', in Burchard Brentjes and Hans-Joachim Peuke, *Wissenschafts-beziehungen zwischen Halle und indien in tradition und Gegenwart*, Halle: Martin-Luther-Universitat Halle-Wittenberg, 1987/44 (142), pp. 27-38.

Chakarabarti, Haripada, 'History of Irrigation in Ancient India', in *Proceedings of the Indian History Congress*, 32nd session, Jabalpur, Delhi: Indian History Congress, 1971, pp. 150-7.

Chang, Hasok, 'Spirit, Air, and Quicksilver: The Search for the "Real" Scale of Temperature', in *Historical Studies in the Physical and Biological Sciences*, vol. 31, no. 2, 2001, pp. 249-84.

Charpentier, Jarl, 'A Treatise on Hindu Cosmography from the Seventeenth Century', in *Bulletin of the School of Oriental Studies*, vol. 3, no. 2, 1924, pp. 317-42.

Dikshitar, V.R.R., 'A History of Irrigation in South India', in *Journal of Indian Culture*, 1945, pp. 71-80.

Grove, H, Richard, 'The Great El Nino of 1789-93 and its Global Consequences Reconstructing Extreme Climate Event in World Environmental History', in *The Medieval History Journal*, vol. X, 2007, pp. 75-98.

Gurukkal, Rajan, 'Aspects of the Reservoir System of Irrigation in the Early Pandya State', in *Studies in History*, vol. II, no. 2, 1986, pp. 155-62.

——, 'Forms and Production of Forces of Change in Ancient Tamil Society', in *Studies in History*, vol. V, no. 2, 1989, pp. 164-5.

Hippocrates, 'Air, Water & Places', in G.E.R Lyold, ed., *Hippocratic Writings*, London: Allen, 1978, pp. 12-16, 148-69.

Karashima, Noboru, 'Allur and Isanamangalam: Two South Indian Villages of the Cola Times', in *Proceedings of the First International Conference-Seminar of Tamil Studies*, Kuala Lumpur: International Association of Tamil Research, 1966, pp. 426-36.

Ladurie, Emmanuel Le Roy, 'Histoire et Climat', *Annales: économies, sociétés, civilisations*, vol. 14, 1959, pp. 3-34.

Martineau, Alfred, 'Les cyclones â la cote Coromandel de 1681 â 1916', *Revue Historique de l'Inde Française*, vol. II, 1917, pp. 229-324.

Naidu, B.V. Narayanasamy, 'Famines in the City of Madras', *Madras Tercentenary Volume*, London: Madras Tercentenary Committee, 1939, pp. 73-87.

Ortiz, M., and R. Bilham, 'Source and Rupture Parameters of the 31 December 1881 Mw" 7.9 Car Nicobar Earthquake Estimated from Tsunamis Recorded in the Bay of Bengal', in *Journal of Geophysical Research*, 2003, pp. 108-221.

Sastri, K.A. Nilakanta, 'Takuapa and its Tamil Inscription', in *Journal of the Malaysian Branch of the Royal Asiatic Society*, vol. XXII, 1949, pp. 25-30.

Srivastava N.N., and K. Ramachandram, 'A New Catalogue of Earthquakes for Peninsular India during 1839-1900', *Mausam*, 36 (30), 1985, pp. 351-8.

Stein, Burton, 'Circulation and Historical Geography of Tamil Country', in *Journal of Asian Studies*, vol. XXXVII, no. 1, November 1977, pp. 7-26.

Stephen, S. Jeyaseela, 'Most Buried Treasure: Letters of the Portuguese Jesuits from Tamil Country in South India during the Seventeenth Century', in *Hispanic Horizon*, vol. XVIII, no. 20, 2001, pp. 76-84.

——, 'Rainfall and Irrigation in Medieval South India and the Socio-Economic Impact', *Proceedings of the Indian History Congress*, 66th Session, Delhi, Indian History Congress, 2006, pp. 296-308.

——, 'Water Resources and Management in the Rural Setting of Medieval South India', in Peter Borschberg and Martin Krieger, eds., *Water and State in Europe and Asia*, Delhi, Manohar, 2008, pp. 105-20.

Shulman, David, 'The Tamil Flood Myths and the Cankam Legend', in Alan Dundes, ed., *The Flood Myth*, Berkeley: California University Press, 1988, pp. 312-14.

Thamby, Siva K., 'Early South Indian Society and Economy: The Tinai Concept', in *Social Scientist*, vol. 29, 1974, pp. 20-37.

Tirumalai, R., 'Allur and Isanamangalam Revisited', in *Svasti Sri: Dr. Chhabra Felicitation Volume*, , Delhi: Agam Kala Prakashan, 1984, pp. 25-55.

——, 'Land Reclamation of Flood Damaged and Sand-Cast Lands: A Study in Price, Rentals and Wages in Later Chola Times', in *Journal of the Epigraphical Society of India*, vol. I, 1984, pp. 64-7.

Walsh, R., R. Glaser and S. Militzer, 'The Climate of Madras During the Eighteenth Century', in *International Journal of Climatology*, vol. 19, issue 9, July 1999, pp. 1025-47.

BOOKS

Alexandre, Pierre, *Le Climat au Moyen Âge en Belgique et dans les Régions Voisines*, Louvain: Centre Belge d'Histoire Rurale, 1976.

——, *Le Climat en Europe au Moyen Âge: Contribution à l'histoire des Variations Climatiques de 1000 à 1425, d'après les sources narratives de l'Europe occidentale*, Paris: Ecole des Hautes Etudes en Sciences Sociales, 1987.

Ambirajan, S., *Classical Political Economy and British Policy in India*, Cambridge: Cambridge University Press, 1978.

Appadorai, A., *Economic Conditions in Southern India (AD 1000-1500)*, 2 vols., Madras: University of Madras, 1936.

Backer, Sommervogel de, *Bibliothèque des Ecriveins de la Compagnie de Jésus*, vol. VI, Paris: Oscar Schepens, 1810.

Bapat, A., R. Kulkarni and S. Guha, *Catalogue of Earthquakes in India and Neighbourhood from Historical Period up to 1979*, Roorkee: Indian Society of Earthquake Technology, 1983.

Braudel, Fernand, *The Mediterranean and the Mediterranean World in the Age of Philip II*, Berkeley: California University Press, 1996.

Cox, Arthur Frederick, *A Manual of the North Arcot District in the Madras Presidency*, Madras: Government Press, 1895.

Des Fosses, Henri Castonnet, *L'Inde Française avant Dupleix*, Paris: Germain & Grassin, 1887.

Dubois A.P., and Xavier Raymond, *Inde*, Paris: Challemal, 1845.

Fenger, Johan Ferdinand, *History of the Tranquebar Mission: Worked out from Original Papers*, trans., Emil Francke, Tranquebar: Evangelical Lutheran Mission Press, 1863.

Francis, W., *Madras District Gazetteers: Madurai*, Madras: Government Press, 1906.

——, *Madras District Gazetteers: South Arcot*, Madras: Government Press, 1906.

Gaebelé, Yvonne Robert, *Une Parisienne aux Indes au XVIIe Siècle: Madame François Martin*, Pondichéry: L'Imprimerie du Gouvernement, 1937.

Grove, H. Richard, *Green Imperialism: Colonial Expansion, Tropical Eden and the Origins of Environmentalism, 1600-1860*, Cambridge: Cambridge University Press, 1995.

——, Damodaran Vinita and Satpal Sangwan, *Nature and the Orient, The Environmental History of South and South East Asia*, Delhi: Oxford University Press, 1998.

Harrison, Mark, *Climates and Constitutions: Health, Race, Environment and British Imperialism in India, 1600–1850*, Delhi: Oxford University Press, 1999.

Heitzman, James, *Gifts of Power: Lordship in an Early Indian State*, Delhi: Oxford University Press, 1998.

Hemingway, F.R.R.B., *Madras District Gazetteers: Tanjore*, Madras: Government Press, 1906.

——, *Madras District Gazetteers: Trichnopoly*, Madras: Government Press, 1907.

Kamal, S.M., *Sethupathi Mannar Varalaaru*, Ramanathapuram: Sharmila Pathippagam, 2003.

Karashima, Noboru, *South Indian History and Society: Studies from Inscriptions, AD 850-1800*, Delhi: Oxford University Press, 1984.

Ladurie, Emmanuel Le Roy, *Times of Feast, Times of Famine: A History of Climate since the Year 1000*, trans., Barbara Bray, London: George Allen & Unwin, 1971.

Launay, Adrian, *Histoire des Missions de l'Inde: Pondichery, Maissour, Coimbatour*, 5 vols., Paris: Societe des Missions Etrangeres, 1895-8.

Love, Henry Davidson, *Vestiges of Old Madras: Traced from the East India Company's Records Preserved at Fort St. George and the India Office and from Other Sources*, 3 vols., London: 1913, reprint, Delhi: Asian Educational Services,1996.

Luc, Jean-André De, *Recherches sur les modifications de l'atmosphère*, 2 vols., Geneva, UNESCO, 1772.

Maclean, C.D., *Manual of Administration of Madras Presidency*, Madras: Government Press, 1885.

Mahalingam, T.V., *Administration and Social Life under Vijayanagara*, Madras: University of Madras, 1940.

Malleson, G.B., *History of the French in India: From the Founding of Pondicherry in 1674 to the Capture of that Place in 1761*, London: Longmans, Green & Co, 1909.

Mazière, Pierre M. la, *Lally Tollendal*, Paris: Plon, 1931.

Minakshi, C., *Administration and Social Life under the Pallavas*, Madras: Univeristy of Madras, 1958.

Morgan, F.E., *Irrigation in Southern India*, London: Foreign and Colonial Compiling Publication Company, 1915.

Pfister, Christian, *Agrarkonjunktur und Witterungsverlauf im Westlichen Schweizer Mittelland, 1755-1797*, Bern: Geographisten Institut der Universitat Bern, 1975.

——, *Das Klima der Schweiz von 1525-1860 und seine Bedeutung in der Geschichte von Bevölkerung und Landwirtschaft*, Bern: Haupt Bern, 1984.

——, *Wetternachhersage: 500 Jahre Klimavariationen und Naturkatastrophen (1496-1995)*, Bern: Haupt Bern, 1999.

Philip, Kavita, *Civilizing Natures: Race, Resources, and Modernity in Colonial South India*, New Brunswick: Rutgers University Press, 2004.

Rajalakshmi, R., *Tamil Polity*, Madurai: Ennes Publication, 1983.

Rajamanickam, M., *Pallavar Varalaru*, Madras: South India Saiva Siddhanta Pathippu Kazhagam, 1976.

Ray, Aniruddha, *The Merchant and the State: The French in India, 1666-1739*, 2 vols., Delhi: Munshiram, 2004.

Raychaudhuri, Tapan, *Jan Company in Coromandel, 1605-90: A Study in the Interrelations of European Commerce and Traditional Economies*, The Hague: Koninklijk Instituut voor Taal-,Land- en Volkenkunde, 1962.

Royer, G.E., *An Economic History of the English Poor Law, 1750-1850*, Cambridge: Cambridge University Press, 1990.

Sastri, Kuppusami, ed., *Srimad Valmiki Ramayanam*, Madras, Vijaya Press, 1958.

Sastri, K.A. Nilakanta, *Studies in Cola History and Administration*, Madras: University of Madras, 1932.

——, *The Colas*, Madras: University of Madras, 1937.

——, *A History of South India from Pre-Historic Times to the Fall of Vijayanagar*, Oxford University Press,1976.

Smith, N., *A History of Dam*, London: Orient Longman, 1971.

Srinivasan, T.M., *Irrigation and Water Supply: South India 200 BC–AD1600*, Chennai: New Era Publications, 1968.

Stein, Burton, *Peasant State and Society in Medieval South India*, Delhi: Oxford University Press, 1980.

Stephen, S. Jeyaseela, *The Coromandel Coast and its Hinterland: Economy, Society and Political System, 1500-1600*, New Delhi: Manohar Publishers, 1997.

——, *Portuguese in the Tamil Coast: Historical Explorations in Commerce and Culture, 1507-1749*, Pondicherry: Navajothi Publishers, 1998.

——, *Kaalaniya Thodakka Kaalam, 1500-1800,* Chennai: NCBH, 2018.

——, *Kaalaniya Valarchi Kaalam, Pulampeyarnthavargalin Vaazhai*, Chennai: NCBH, 2018.

——, *Subaltern Lives: Tamil History and Society from Medieval to Early Modern Age*, Delhi: Kalpaz Publications, 2019.

——, *Goodbye to Tamil Motherland: The Rise of Labour Migration Overseas and the Society, 1729-1890*, Delhi: GenNext Publication, 2019.

Subbarayulu, Y., *Studies in Cola History*, Chennai: Oxford University Press, 2001.

Subramanian, N., *Pre-Pallavan Tamil Index*, Madras: Ennes Publications, 1990.

Thornton, Edward, *The History of the British Empire in India*, London: Longmans, 1859.

Venkatramayya, K.M., *Administration and Social Life under the Maratha Rulers of Thanjavur*, Thanjavur: Tamil University, 1984.

Venkataramanaya, N., *Studies in the History of the Third Dynasty of Vijayanagara*, Madras: University of Madras, 1939.

Vigié, Marc, *Dupleix*, Paris: Fayard, 1993.

Weber, Jacques, *Les Etablissements francais en Inde au XIX e Siecle (1816-1914)*, 5 vols., Paris: Librairie de l'Inde, 1988.

Wheeler, J. Talboys, *Early Records of British India: A History of the English Settlements in India*, Calcutta: Government Press, 1879.

DISSERTATIONS

Gurukkal, Rajan, *The Agrarian System and Socio-Political Organization Under the Early Pandyas, 600-1000*, PhD Dissertation, New Delhi: Jawaharlal Nehru University, 1984.

Suresh, B., *Historical and Cultural Geography and Ethnography of South India from the Chola Records*, PhD Dissertation, Poona: Deccan College, 1965.

Index

aaru 23, 26
aatrukulai 56
aayar 13
abishekham 84
Achutappa nayak 36-7
Achyutadevaraya 74
Adyar 103
Aganaanuru 98
agriculture 14, 23, 30
Akkalimangalam 37
Alagarkoil 32
Alangudi 69, 71, 73
Alexandre, Pierre 174
Allur 127
Anaimalai 27
Andaman Islands 139
Andhra Pradesh 16
Arakandanallur 69
Arasalar 113
arasaperru 23
Arcot 83
Ariksesarinallur 29
Armagon 99
Aruppukottai 30
Asylum Press 52
atmospheric pressure 145
Attrukaal vetti 56
Attrupaatam 56
Ayyampettai 69

Bahur 27-8
Balfour, Gibson 181
Barbier, Claude 79
barometer 121, 147, 153, 157
battle of Talikota 74
Baulez, Marius J. 80
Bay of Bengal 26, 98
Bayley 45
Bellary 16
Bentinck, William 85
Berhampur 16
Bhaiyappa nayak 35
Blanford, Henry F. 46, 47, 50-1, 54, 120, 167
Blunt thermometer 19
Brahamadeya 31
Brahmins 28-9, 84
Braudel, Fernand 174

canals 26
Carter 99
Celsius, Anders 148
Chambers, William 138
Chandragiri 35, 145
chatty 42
Cheyyar 36
Chidambaram 26, 138
Chingleput 36, 52, 68, 85
Chinnakollaipatti 29, 33
Chittoor 36
Cholamandala Sathakam 24
Chunnambar 134
Cleghorn, Hugh 181
Cogan 99
Coimbatore 19, 25, 50-1
Collet, Joseph 161
Collingwood, R.G. 16
Colonel James Capper 19, 150
Colonel Smith 45
Coonoor 164
Cotton, Sir Arthur 26
Cox 56

Cuddalore 18-9, 54, 84, 102, 106, 112, 131, 136-7
Cyclone 41, 98-100

Dalavanur 32
Dalavaypuram 34
Dam 23-4
Danish Halle mission 38
dasavanda 35
devadana 31
Devikapuram 35
Dhurjati, 73
Dindigul 27
Dodabetta 159, 169
drought 67
Dupleix 80, 107
Durgarajapattinam 99
Dutch 16-17

earthquakes 127, 139-40
El Nino 176
Ellappa nayak 35
Elphinstone Bridge 120
Emanuel de Veiga 145
Emmanuel Le Roy Ladurie 174
Erumbur 69, 73
Etti Saathan 33
eyinar 13

Fahrenheit 148, 153, 159
famine 67-80
fishermen 112
fishing boats 112
floods 127-30
Fort St. David 132
Fort St. George 46, 100
Francisco Macedo 75
Francois Jean-Marie Laouenan 80, 90
Francois Martin 78, 107
Franklin, J.J. 18, 117, 119, 120-1
French Guyana 91

Gadilam 137
gales 114-5
Ganjam 16
gardens 14, 105
Geister, Johann Ernest 18, 38
German 17
Gilchrist, William 19, 160
Gingee 36
Godavari 16
Goldingham, John 18-9, 43-5, 114, 117, 150
great famine 18, 87-90
Guadeloupe 91
guns 106

Halley, Edmund 59
Hill, Henry Green 99
historical geography 13
Hosur 140
Hunter 46, 48, 53
hurricane 101, 119
hygrometer 152-3

idaiyar 13
Ilangokudi 35
Ilayangudi 67, 73
Iluppaikudi 29
inamdars 55
irai kaval 28
Iraiyanar Ahapporul 68
Iraiyur 37
irandadi kinaru 34
Irrigation Cess Act 55
Isanamangalam 127

jala-sutrada 36
Jambai 73
Jambhu Thivu 15
Jambukeswaram temple 37
Jean Fourcade 81
Jean François Charpentier de Cossigny 131
Jesse Ramsden barometer 19, 149
Jesuits 25, 76, 79, 145
John, Christoph Samuel 105
Joseph Louis Gay-Lussac 148

K.V., Naidu 55
kaaduvettis 23
Kalahasti 73
kalams 23
kalavar 13
Kaliyan yeri 27
kallanai 24
Kambar 34
Kanavanakkari 34
kanavar 13
Kanchipuram 67
kanikkai 23
Kanyakumari 14, 32
Kappalur 30
Karaikal 18, 32, 105-6, 136
karais 22, 24
Karnataka 16
kartthumpu 56
Kasakudi plates 34
kasu 15, 80
Kavatapuram 15
Kaveri 14, 23, 50, 127
Kaveri Kamukha 14
Kaviri 14
Kaviripoompattinam 15
Kerala 15
Keralantaka Viluppariayan 24
Kilavan Sethupathi 37
Killiyur 30
kinaru 34-5
King of Mysore 25-6
Klein, Jacob 103
Kodangi 36
Kodungalur 36
Kollidam 26, 127, 132
Kopperunjika I 14
Kudimaraamath 54
kulam 22, 33
Kulothunga Chola III 31, 129
Kumbakonam 24
kunavar 13
Kunnathur 33
Kuntalur 24
kurinji 13
Kuvam 101

Le Moniteur de La Flotte 89
Lieutenant Henry Kater 153
Lingamayya nayak 36

maa 15
Madhuranthakam 56
Madras 16, 18, 38, 54, 83, 111-2, 130
Madras Journal of Literature and Science 52
Madras Observatory 19, 45, 150, 159
Madras Presidency 54-5
Madurai 15, 19, 25, 30, 50-1, 54, 82
Mahabalipuram 138
mahasabha 29, 129
Mahendravadi kulam 56-7
Major Worster 45
Malattar 137
Mallappa nayak 36
Mangammal 26
manigramam 99
Manimekalai 15
Manimukta river 138
Maniyakkarar Choultry 76
Manjakuppam 136
Maran yeri 28
Maranaur 29
maravar 13
Maravarman Srivallabha 129
Martin, Pierre 25
Marudhan Ilanaaganaar 98
marutham 13
Maruthur 27
mathagu 29, 30, 32
Mauritius 91
meenavar 13
Metallic Tube Barometer 19
mirasdar 55
Monsieur Floyer 106
mullai 13
Mullinadu 35

Musiri 31
Muthugulathur 37
Muthur 15
Mylapore 18, 74, 100-1

Naaladi kinaru 34
Naavalam 14
Nagapattinam 18-9, 75, 91, 105, 112, 130
Nagore 150, 168
Nairne thermometer 19
Nandalaru 132
Nandivarman II 14, 34
Nanguneri 30, 32
Nannilam 24
Narasimha Varman II 67
Narasingamangalam 27
nattars 30, 71
Nayinappa nayak 35
Nedungunram 36
Neer aarambam 56
Neer aaru 56
Neer amanji 56
Neer nilai 56
Neer vilai kuli 56
Neerk kirai 56
Neerkiya vilai neer kuli 56
Neernilai Kaasu 56
Nellikuppam 36
Nemili 28
Nerkunram 31
Nettaipakkam 129
Nicobar Island 139
Nicolau Manucci 139
Nilaiyur 33
Nilakanta Sastri, K.A. 13
Nilgiris 19, 167-8
nirkuli 36
North Pennar 50

Odisha 16
Olukkai 31
Ootacamund 164-5

paalai 13
Padaividu 36
Palar 23, 50, 72, 127
Palayamkottai 50, 52
Pallavarayan 32
Pallavas 23, 55
Palli kulam 28
Pallichandal 73
panam 23, 36, 76, 78
Pandyas 15, 23, 30-1
Panjai 69
Paraiyar 75
Paramakudi 37
Parankusapputtur 29
parathavar 13
Paripaadal 22
pathakku 56
Pathittru Paththu 15
Penang 91
Pennar 23
Periyanattar 30
Perumbakkam 37
peryeri 28
Pfister, Christian 174
picottah 24, 34
Pillai, Ananda Ranga 40, 107-8, 122, 178
Pillai, Muthu Vijaya Thiruvengadam 40, 58
Pillai, Rangappa Thiruvengadam 40
Pillai, Sandira 89
Pillai, Subburaya 89
Poissevin 80
Pondicherry 18-9, 27, 40-1, 77-8, 107, 130-1
Ponnaiyar river 137
Ponnu Thambi Pillai 90
Poonamalee 140
Porto Novo 18, 100, 106-7
Pothuau, Admiral 89
prasaatham 36
Proenca, Antao 76
Public Works Department 54, 86-7

Pudukottai 27, 31
Pulicat 18, 76, 101, 112
Pullan yeri 27
Punganur 27
Purananuru 22
Puthupaadi 128

Rain gauge 42, 45
Rainfall 18, 22, 38-43
Rainfall statistics 45-7
Raja Raja III 31
Raja Simha 67
Rajendra Chola I 127
Ramanathapuram 27, 33, 78, 83
Ramappa nayak 35
Ramayana 14
Réaumur 148
Reddi, Bommu 37
Regnault, Henri Victor 148
revetment 28
Roxburgh, William 19, 42, 149, 176

sabha 27, 34, 57, 71
Sadurangapattinam 18, 102, 109-10, 112, 119
Saidapettai 103
Saiyam hills 14
Salem 85, 140
Saluva Thirumalai Devaraya 31
Samburvarayas 23
sammadam 23
Sangam literature 13, 15, 22
satantaka 36
Sattur 28-9, 33
Self-registering Barometer 19
Sembiyanallur 37
seneer poduirai 56
seneer vetti 56
sennir 35
Sethupathi ruler 35
Sethupathi, Muthu Ramalinga Vijaya Raghunatha 37
Sevappa nayak 36
Shahji II 26
Shenkottai 50, 52
Shervaroy Hills 140
Sholingur 26
Siddhalingamadam 30
Silapathikaram 15, 24
silasasanam 36
Singapore 91
Sir Thomas Herbert 133
Sirkazhi 15, 24
slaves 24, 70, 75-6
sluice 29
Somangalam 128
South Pennar 31, 50
Sri Kalahasti Mahatmyam 73
Srimara Sri Vallabha 32
Sriperumbudur 140
Sriranga nayak 35
Srirangam 127
Srivilliputtur 29
St. Thomas Mount 150
Stein, Burton 13, 37
Stevenson, William 146
sthanattar 36
storm 41, 43, 44
Strachey, Richard 46, 49, 52, 88
Suchindram 31-2
Sundara Pandya Kandiyadevan 30
Suralur 25
Sympiesometer 121

Takua-pa 98
Taylor, Thomas Glanville 158
Terra Tamilica 14
Thailand 98-9
Thamiraparani 23
Thanjavur 19, 25, 31, 80, 83, 105, 132-3
thinais 13
Thirukkanchi 28
Thirukkural 14, 22, 98
Thirumangalakkudi 69
Thiruppachuram 68

Thiruvalluvar 14, 22, 98
Thiruvanaikaval 127
Thiruvayppadi 31
Thiruverumbur 30
Thomas Bowrey 38
Thomas Munro 85
thotti 34
thozhuvar 13
thumpu 33
Timmanankuppam 36
Tirubhuvanai 28
Tiruchirapalli 14, 19, 24, 25, 31, 50, 82
Tirukachchur 68-9, 71, 73
Tirukarugavur 68-9
Tirukkadaiyur 35, 69
Tirukkoyilur 136
Tirumalai nayak 35
Tirumalarayanpattinam 32
Tirunelveli 33
Tirupapuliyur 106, 136
Tirupati 35
Tiruppanangadu 69
Tiruvaalangadu 31
Tiruvadi 71
Tiruvaduturai 24
Tiruvamathur 37
Tiruvannamalai 69, 70, 72
Tiruvarur 105
Tiruvaththur 15
Tiruvenkadu 24
Tiruvottur 128
Tome Pires 99
Tranquebar 18, 19, 104-5, 130-1
tributaries 26
Trillard, Adolph 89-90
Triplicane 101
Tsunamis 127, 138-9

Udaya Chandran 14
Ukkira Pandyan 15
ulliyak-kuli 34
Upparu river 108
Uttiramerur 26-7
Uyyakondan vaykkal 31
uzhavar 13

Vaigai 14, 23, 30, 32, 37, 50
Vairameghatataka 26
Vajara Bhoti 67
valayar 13
Valuvur 128-9
Vandavasi 140
Vandipalayam 106, 132
varigripakarana 37
Vayiramega Chaturvedimangalam 28
vaykaals 29, 30-1
Vedampattu 37
Vegavati 14
Velan Paranjothi 29
veli 23
Vellar 50
Velur Palayam 68
vettaparru 28
Vijayamangalam 33
Vijayanagara 23
Vijayanarayanam 31, 128
Vikrama Chola 129
village headmen 17
Villiyanoor 131
Villupuram 136, 140
Vira Pandya 71
vittavan vilukaadu 36
Vizagapattinam 103
Voelcker, John Augustus 88
Vriddhachalam 138

water cess 36
water channels 17
water disputes 35
water management 17
water rights 35
water tanks 17
Wellington 164, 166-7
wells 17
Wright, Robert 181

Yerelupathu 34
yeri 23, 26
yeri ayam 28, 56
Yeri meen kaasu 56
Yeri meen paatam 56
Yeri neer kovai 56
Yeri patti 27, 56
yeri vaariyam 27
Yeri vaay padikkaval 56
yerudoovaal pullu 154
yettam 34

zamindar 55